Reading Writing Interfaces

Electronic Mediations

Series Editors: N. Katherine Hayles and Samuel Weber
Founding Editor: Mark Poster

44 *Reading Writing Interfaces: From the Digital to the Bookbound*
Lori Emerson

43 *Nauman Reiterated*
Janet Kraynak

42 *Comparative Textual Media: Transforming the Humanities in the Postprint Era*
N. Katherine Hayles and Jessica Pressman, Editors

41 *Off the Network: Disrupting the Digital World*
Ulises Ali Mejias

40 *Summa Technologiae*
Stanisław Lem

39 *Digital Memory and the Archive*
Wolfgang Ernst

38 *How to Do Things with Videogames*
Ian Bogost

37 *Noise Channels: Glitch and Error in Digital Culture*
Peter Krapp

36 *Gameplay Mode: War, Simulation, and Technoculture*
Patrick Crogan

35 *Digital Art and Meaning: Reading Kinetic Poetry, Text Machines, Mapping Art, and Interactive Installations*
Roberto Simanowski

34 *Vilém Flusser: An Introduction*
Anke Finger, Rainer Guldin, and Gustavo Bernardo

33 *Does Writing Have a Future?*
Vilém Flusser

32 *Into the Universe of Technical Images*
Vilém Flusser

31 *Hypertext and the Female Imaginary*
Jaishree K. Odin

30 *Screens: Viewing Media Installation Art*
Kate Mondloch

(continued on page 224)

Reading Writing Interfaces

From the Digital to the Bookbound

Lori Emerson

Electronic Mediations 44

 University of Minnesota Press
Minneapolis • London

The University of Minnesota Press gratefully acknowledges financial assistance provided for the publication of this book from the Eugene M. Kayden Research Grant through the College of Arts and Sciences at the University of Colorado at Boulder.

Every effort was made to obtain permission to reproduce material in this book. If any proper acknowledgment has not been included here, we encourage copyright holders to notify the publisher.

Portions of chapter 4 were previously published as "My Digital Dickinson," *The Emily Dickinson Journal* 17, no. 2 (2008): 55–76. Copyright 2008 by The Johns Hopkins University Press. Reprinted by permission.

Copyright 2014 by the Regents of the University of Minnesota

Published by the University of Minnesota Press
111 Third Avenue South, Suite 290
Minneapolis, MN 55401–2520
http://www.upress.umn.edu

Library of Congress Cataloging-in-Publication Data
Emerson, Lori.
    Reading writing interfaces : from the digital to the bookbound /
Lori Emerson.
(Electronic mediations ; 44)
    Includes bibliographical references and index.
    ISBN 978-0-8166-9125-8 (hc : alk. paper)
    ISBN 978-0-8166-9126-5 (pb : alk. paper)
1. Hypertext literature—History and criticism. 2. Literature and technology. I. Title.
    PN56.I64E3625   2014
    802'.85—dc23
                                                    2013038695

Printed in the United States of America on acid-free paper

The University of Minnesota is an equal-opportunity educator and employer.

20 19 18 17 16 15 14   10 9 8 7 6 5 4 3 2 1

Contents

Acknowledgments   vii

Introduction: Opening Closings   ix

1. Indistinguishable from Magic: Invisible   1
   Interfaces and Digital Literature as Demystifier

2. From the Philosophy of the Open to the   47
   Ideology of the User-Friendly

3. Typewriter Concrete Poetry as Activist   87
   Media Poetics

4. The Fascicle as Process and Product   129

Postscript: The Googlization of Literature   163

Notes   185

Index   219

Acknowledgments

So many people helped this book find the light of day. To name just a few, my sincere gratitude goes to Mark Amerika, Doug Armato, Derek Beaulieu, Kathi Inman Berens, bill bissett, Brett Bobley, Clint Burnham, J. R. Carpenter, Judith Copithorn, Frank Davey, Craig Dworkin, Christopher Funkhouser, David Glimp, Kenneth Goldsmith, Jeremy Green, Dene Grigar, Katherine Harris, Susan Howe, Geof Huth, Danielle Kasprzak, Matthew Kirschenbaum, William Kuskin, Deena Larsen, Marjorie Luesebrink, Mark Marino, Cristanne Miller, Nick Montfort, Helmut Mueller-Sievers, Eleanor Nichol, Jussi Parikka, Jasia Reichardt, Jason Rhody, Tim Roberts, Marie-Laure Ryan, Marvin Sackner, Stephanie Strickland, Joseph Tabbi, Darren Wershler, and Paul Zelevansky. I often shamelessly used Twitter to think out loud as I wrote, and I thank @eetempleton, @flangy, @giallo, @hedorah55, @hyperverses, @joshhonn, @karikraus, @Leonardo_UPRM, @mjntendency, @ncecire, @noeljackson, @pbenzon, and @spold for the support, the nudges, and the stimulating conversation. I owe thanks to my dear friends Doreen Martinez and Randy Prunty and my other dear friends on the GS Boulder Cycling Team who kept me sane, healthy, and happy—especially Lorna Pomeroy Adley, Virginia Betty, Natasha Kelly, and Ann Remmers. Finally, I am quite sure this book would not exist at all if not for the tireless encouragement from my mom and stepdad and, far above and beyond all, from Benjamin Robertson, who was nobly unfailing in his willingness to support the project and listen, read, and respond to every word I wrote.

Opening Closings

Interface

This book begins and ends with magic—sleights of hand that disguise how closed our devices are by cleverly diverting our attention to seemingly breathtaking technological feats. From the stylized, David Copperfield–inspired Apple launch for the iPad, which is touted as a "truly magical and revolutionary product," to (as of this writing) the impending launch of Google Glass, which is already being marketed as a device that will provide "answers without having to ask," we are well into the era of the marvelous. It's marvelous in the sense of that which is wondrous—for how could we not wonder at how the iPad simulates, for example, the relationship between inertia and friction or at how Google Glass is an invisible portal to information now embedded into our very perception of the world? But it's also marvelous in the sense that these devices seem to have supernatural properties. But, of course, supernatural they are not.

Reading Writing Interfaces is, then, anything but a breathless account of the wonders of contemporary digital computing. From beginning to end, it is about demystifying devices—especially *writerly* demystification—by opening up how exactly interfaces limit and create certain creative possibilities. Technological constraints are nothing new, for—as I discuss in chapter 4—Emily Dickinson's work with fascicles clearly records her fine-tuned understanding of pen, pencil, paper, and even pinning as interface. But what is new is that the interfaces themselves and therefore their constraints are becoming ever more difficult to perceive because of the blinding seduction of the wondrous that at least partly comes back into view again once we undertake an excavation of how things (could) have been otherwise.

While *interface* is a productively open-ended, cross-disciplinary term, generally speaking in computing it refers simply to the point of interaction between any combination of hardware/software components. Florian Cramer has, however, usefully delineated eight different kinds of interface, including hardware-to-hardware interfaces (such as sockets and drives), hardware controllers for software functions (such as joysticks), software-to-hardware interfaces (such as the operating system), and—especially relevant for this book—human-to-hardware interfaces (such as keyboards, screens, and mice) and human-to-software interfaces (such as the graphical user interface [GUI]).[1] Throughout, however, I settle on an even more expansive definition so that interface is a technology—whether it is a fascicle, a typewriter, a command line, or a GUI—that mediates between reader and the surface-level, human-authored writing, as well as, in the case of digital devices, the machine-based writing taking place below the gloss of the surface.[2] The interface is, then, a threshold, but in a more complex sense than simply that which opens up from one distinct space to another distinct space. Instead, I draw on Alexander Galloway's articulation of interface as "the point of transition between different mediatic layers within any nested system" as a way to highlight the fact that while interface does grant access, it also inevitably acts as a kind of magician's cape, continually revealing (mediatic layers, bits of information, etc.) through concealing and concealing as it reveals.[3] With the advent of so-called interface-free devices such as Google Glass and the iPad—"interface-free" in the sense that, as multitouch designer Jeff Han enthused in 2006, "there's no instruction manual" and in the sense of Apple's and Google's favorite marketing slogan of the moment, "It just works"—largely what's at issue in this book is what's revealed, or what writers in particular reveal via practices of media poetics, through what is concealed. The dream in which the boundary between human and information is eradicated is just that—a dream the computing industry rides on as it attempts to convince us that the dream is now reality through sophisticated sleights of hand that take place at the level of interface.

Throughout, I identify these interfaces that obscure ever more from the user in the name of "invisibility" and the "user-friendly" with what's fast becoming an ideology. I use *ideology* not merely in the sense of the adamant belief in making the computer more approachable but more in the sense that *user-friendly* is used quite deliberately to distort reality by convincing users that this very particular notion of a user-friendly device—one that depends on and then celebrates the device as entirely closed off both to the user and to any understanding of it via a glossy interface—is the only possible version of the user-friendly, one that claims to successfully bridge the gap between human and computer. In reality, the glossy surface of the interface further alienates the user from having access to the underlying workings of the device.

I am not arguing wholesale against user-friendly interfaces that seek to be invisible, as long as user-friendliness and usability, on the one hand, and creativity, tinkering, and making, on the other hand, are not mutually exclusive. Even invisibility has its place in interface design, for the consistency of contemporary typing interfaces produces familiarity and, in turn, a kind of invisibility that is precisely what enables me to type quickly and efficiently without looking at my fingers or enables me to easily create, organize, and save this word-processing document via the metaphor of the desktop. Who would want an interface that constantly and intentionally glitches, fails, and disrupts? Janet Murray is, then, in some sense right to declare that designers should focus on transparency, for "a good interface should not call attention to itself, but should let us direct our attention to the task."[4] But when transparency not only transforms into that which is valued above all else but also becomes an overriding, unquestioned necessity, it turns all computing devices into appliances for the consumption of content instead of multifunctional, generative devices for reading as well as writing or producing content. Galloway is, however, the necessary counterpoint to Janet Murray, for where she urges designers to achieve transparency, Galloway declares that "for every moment of virtuosic immersion and connectivity, for every moment . . . of inopacity, the [interface as] threshold becomes one notch more invisible, one notch more inoperable. . . . Operability engenders inoperability."[5] The foregoing is, oddly, both rhetorical exaggeration and accurate description, for as long as I can continue to type ninety words per minute or effortlessly organize my word-processing documents, in itself an achievement of habit made possible by interface design, I cannot also understand or intervene in the underlying workings of either one. These closed computing interfaces that are well on their way toward invisibility are both operable and inoperable, the one at the cost of the other.

Media Archaeology, Media Poetics

Friedrich Kittler infamously writes, "Media determine our situation, which—in spite or because of it—deserves a description."[6] Despite our best efforts to literally and figuratively bring these invisible interfaces back into view, either because we are so enmeshed in these media or because the very definition of ideology is that which we are not aware of, at best we may only partly see the shape of contemporary computing devices. Media archaeology (with Kittler as one of its deep influences) provides, however, a sobering conceptual friction in the way that certain theorists identified with the field, such as Geert Lovink, use it to undertake "a hermeneutic reading of the 'new' against the grain of the past, rather than telling of the histories of technologies from past to present."[7] (Usually, aside from the more general philosophical problems inherent to any teleology, the result of this model of media history that leads neatly into the present or even the near future is a triumphalist celebration of the way things are rather than a sharpened awareness of the contours of everyday media apparatuses.) On the whole, media archaeology does not seek to reveal the present as an inevitable consequence of the past but instead looks to describe it as one possibility generated out of a heterogeneous past.[8] Also at the heart of media archaeology is an ongoing struggle to keep alive what Siegfried Zielinski calls *variantology*—the discovery of "individual variations" in the use or abuse of media, especially those variations that defy the ever-increasing trend toward "standardization and uniformity among the competing electronic and digital technologies." Following Zielinski, I uncover a nonlinear and nonteleological series of media phenomena—or ruptures—as a way to avoid reinstating a model of media history that tends toward narratives of progress and generally ignores neglected, failed, or dead media. That said, following on the debates in the field of digital humanities about the connection of theory and praxis (the so-called more hack, less yack debate), this book is more about *doing through thinking* than theorizing media archaeology. To borrow from Jussi Parikka's *What Is Media Archaeology?*, my book "thinks" media archaeologically as the focus on different reading/writing interfaces shifts back and forth (within each chapter and also from one chapter to the next) from the present to the past and back to the present again, all the while interweaving analyses of writing from contemporary digital authors such as Jason Nelson and Young-Hae Chang Heavy Industries, back to Deena Larsen, Paul Zelevansky, and bpNichol, further back to Emily Dickinson, and returning to the contemporary once again via Darren Wershler/Bill Kennedy, Tan Lin, and John Cayley/Daniel C. Howe. These writers all work with and against interfaces across various digital and analog media to undermine not only normative reading/writing practices but, above all, the assumed transparency of conventional reading and writing interfaces.

While I discuss these writers in relation to digital literature and the more institutionally accepted "electronic literature," throughout I also touch on bookbound poetry and digital poetry, for poets especially have long been attuned to—even written through—the distinct material limits and possibilities of writing interfaces of all kinds. But what I have also found is that when writers read or even record their writing interfaces, *through* writing, the result is necessarily a highly visual, tactile literary object that corresponds to traditional literary genres such as poetry or fiction only to the extent that the author names their work as such. As I show in chapter 3, typewriter- and copier machine–based concrete poetry from the 1960s and the 1970s clearly takes part in what I call *media poetics*—the literary exemplar of media archaeology and a practice that extends deep from within the analog and well into the digital. More, while media poetics may erode traditional literary genres such that poetry could be visual art as much as it could be fiction and vice versa, now that we are all constantly connected to networks, driven by the new invisible—formidable algorithms—media poetics is fast becoming a practice not just of experimenting with the limits and possibilities of writing interfaces but rather of *readingwriting*: the practice of writing through the network, which as it tracks, indexes, and algorithmizes every click and every bit of text we enter into the network, is itself constantly reading our writing and writing our reading. As I conclude in the postscript, this strange blurring of, even feedback loop between, reading and writing signals a definitive shift in the nature and definition of literature.

Finally, while *Reading Writing Interfaces* emerges from the conjunction of thinking through doing media archaeology, media poetics, and my own archival research, it's also evidence of doing through building, curating, and tinkering in the Media Archaeology Lab (MAL) at the University of Colorado at Boulder—a lab that I founded in 2009 for cross-disciplinary experimental research and teaching that features the active use of hardware, software, platforms, and tools of all kinds from the past.[9] Without having the ability to directly discover what one might call the variantology of early computing, without experiencing what it's like to operate a computer that predates standardized interfaces (working through, for example, the nonobvious differences between a Commodore key and open-Apple and closed-Apple keys) and whose target audience is the DIYer, the tinkerer, the curious, I would never have understood to the extent that I do now the (equally nonobvious) ideology of the user-friendly.

The MAL houses most of the computers I discuss throughout, including the Apple II, the Apple Lisa, and the Apple Macintosh, as well as many early works of digital literature. The Apple II and the Lisa are particularly important for understanding the history of personal computing and computer-mediated writing, for the shift in interface from the one to the other—and therefore the shift in the limits and possibilities for what one could create—is remarkable. The Apple II series of computers all used the command-line interface, and they were also the first affordable, user-friendly personal computers and, therefore, were the most popular, while the Apple Lisa was the first commercial computer to use a GUI. In terms of the literature produced on these machines, as I write in chapter 2, a work such as *First Screening* by bpNichol—created in 1983–84 using an Apple IIe and the Apple BASIC programming language—is exemplary in that once one accesses it via the original 5.25-inch floppy on an Apple IIe, one finds that the now widely available media translations of it lack nearly all of the crucial material and multilayered aspects of the original. On the one hand, where would we be if *First Screening* wasn't first recovered by Jim Andrews, Geof Huth, Lionel Kearns, Marko Niemi, and Dan Waber, made available via emulator, HyperCard, and Quick-Time movie, and now preserved on both the U.S.-based Electronic Literature Directory and the European-based ELMCIP Knowledge Base? On the other hand, when reading *First Screening,* there is simply no substitute for the command-line interface paired with the physical structure of the Apple II computer. Everything about the Apple II, its entire hardware and software system, offers both writer and reader an utterly different set of experiences than when they read or write on, say, a MacBook or a PC or when they read *First Screening* by way of a GUI. For example, one would never know from the QuickTime emulation that *First Screening* is a series of poems whose meaning is activated through the writer/programmer's invitation to the reader/user to type in commands—from the fact that you have to type "RUN" to initiate it (and of course there's no instruction to "type RUN") to the fact that in line 110 of its code Nichol writes: "REM FOR THE CURIOUS VIEWER/READER THERE'S AN 'OFF-SCREEN ROMANCE' AT 1748. YOU JUST HAVE TO TUNE IN THE PROGRAMME." As Jim Andrews discovered in the process of creating the emulations, "The poem is off-screen in the sense that to play/view it, you have to type in a command"—either RUN 1748, RUN 1748-, GOSUB 1748, GOSUB 1748-—"you have to engage with the language machine at that level to view the poem that remains off-screen until you summon it."[10]

The MAL is, then, a kind of thinking device that enables us to tinker and to track writing-as-tinkering in early works of digital literature; providing access to the utterly unique material specificity of these computers, their interfaces, their platforms, and their software also makes it possible to defamiliarize or make visible for critique contemporary *invisible* interfaces and platforms.

Overview

The following chapters move from the present moment to the past and back to the present, each using a particular historical moment to understand the present. *Reading Writing Interfaces* begins with digital writers' challenge to the alleged invisibility of ubiquitous computing and multitouch in the early twenty-first century and moves to poets' engagement with the transition from the late 1960s' emphasis on openness and creativity in computing to the 1980s' ideology of the user-friendly GUI, to poetic experiments with the strictures of the typewriter in the 1960s and 1970s, and finally to Emily Dickinson's use of the fascicle as a way to challenge the coherence of the book in the mid- to late nineteenth century. At each point in this nonlinear history, I describe how this lineage of media poetics undermines the prevailing philosophies of particular media ecology and so reveals to us, in our present moment, the contours of our contemporary technologies. By the time I return to the present in the postscript, via the foregoing four techno-literary ruptures, I have made visible a long-standing conflict between those who would deny us access to fundamental tools of creative production and those who work to undermine these foreclosures on creativity. In many ways, then, my book reveals the strong political engagement driving a tradition of experimental writing, and it implicitly argues for the importance of the literary in the digital age.

In the first chapter, "Indistinguishable from Magic: Invisible Interfaces and Digital Literature as Demystifier," I begin by describing contemporary claims about ubiquitous computing (ubicomp) as the definitive technological innovation of this century—claims that consistently tout the wonders of invisible interfaces and how they provide us with a more natural, more direct, inherently better way to interact with our computers. Without attention, however, to the ways in which interfaces are anything but invisible in how they frame what can and cannot be said, the contemporary computing industry will only continue unchecked in its accelerating drive to achieve the perfect black box not only through the latest ubicomp devices but also through parallel developments such as so-called Natural User Interfaces, Organic User Interfaces, and even the now widely prevalent multitouch interfaces. All of these interfaces share a common goal underlying their designs: to efface the interface altogether and so also to efface our ability to read, let alone write, the interface, definitively turning us into consumers rather than producers of content. After I return to original writings on early multitouch platforms such as Myron Krueger's Videoplace and even on ubicomp to show how these were created in an utterly different spirit (one that rejected the value of invisibility), I then describe a growing body of digital literature that embraces *visibility* by courting difficulty, defamiliarization, and glitch and that stands as an antidote to ubicomp and this receding present. Writers ranging from literary app creators Jörg Piringer, Jason Edward Lewis, and Erik Loyer to the web-based Nick Montfort, Deena Larsen, Talan Memmott, Judd Morrissey, Jason Nelson, and Young-Hae Chang Heavy Industries all advance a twenty-first-century media poetics by producing digital texts that are deliberately difficult to navigate or whose interfaces are anything but user-friendly. At the heart, then, of some of the most provocative work of digital literature lies a thoroughgoing engagement with difficulty or even failure. By hacking, breaking, or simply making access to interfaces trying, these writers work against the ways in which these interfaces are becoming increasingly invisible even while these same interfaces increasingly define what and how we read/write.

The second chapter, "From the Philosophy of the Open to the Ideology of the User-Friendly," uncovers the shift from the late 1960s to the early 1980s that made way for the very interfaces touted as utterly invisible that I discuss in chapter 1. Based on work I did in the Media Archaeology Lab using many of the original machines and also driven by original archival research I undertook of historically important computing magazines such as *Byte, Computer,* and *Macworld,* as well as handbooks published by Apple Inc. and Xerox, I bring to light the philosophies driving debates in the tech industry about interface and the consequences of the move from the command-line interface in the early 1980s to the first mainstream windows-based interface introduced by Apple in the mid-1980s. I argue that the move from a philosophy of computing based on a belief in the importance of open and extensible hardware to the broad adoption of the supposedly user-friendly GUI, or the use of a keyboard/screen/mouse in conjunction with windows, fundamentally changed the computing landscape and inaugurated an era in which users have little or no comprehension of the digital computer as a medium. Thus, media poetics prior to the release of the Apple Macintosh in 1984 mostly took the form of experimentation with computers, such as the Apple IIe, that at the time were new to writers. Digital poetry/literature from the early 1980s by bpNichol, Geof Huth, and Paul Zelevansky does not work to make the command-line or Apple IIe interface visible so much as it openly plays with and tentatively tests the parameters of the personal computer as a still-new writing technology. This kind of open experimentation almost entirely disappeared for a number of years as Apple Macintosh's design innovations and their marketing made open computer architecture and the command-line interface obsolete and GUIs pervasive.

In the third chapter, "Typewriter Concrete Poetry as Activist Media Poetics," I delve into the era, from the early 1960s to the mid-1970s, in which poets, working heavily under the influence of Marshall McLuhan and before the widespread adoption of the personal computer, sought to create concrete poetry as a way to experiment with the limits and the possibilities of the typewriter. These poems—particularly those by the Canadian writers bpNichol and Steve McCaffery and the English Benedictine monk Dom Sylvester Houédard—often deliberately court the media noise of the typewriter as a way to draw attention to the typewriter-as-interface. As such, when Andrew Lloyd writes in the 1972 collection *Typewriter Poems*, "A typewriter is a poem. A poem is not a typewriter," he gestures to the ways in which poets enact a media analysis of the typewriter via writing as they cleverly undo stereotypical assumptions about the typewriter itself: a poem written on a typewriter is not merely a series of words delivered via a mechanical writing device, and for that matter, neither is the typewriter merely a mechanical writing device. Instead, these poems express and enact a poetics of the remarkably varied material specificities of the typewriter as a particular kind of mechanical writing interface that necessarily inflects both how and what one writes. Further, since they are about their making as much as they are about their reading/viewing, if we read these concrete poems in relation to Marshall McLuhan's unique pairing of literary studies with media studies—a pairing that is also his unique contribution to media archaeology *avant la lettre*—we can again reimagine formally experimental poetry and poetics as engaged with media studies and even with hacking reading/writing interfaces. More, this chapter also draws on archival research not only to uncover the influence of McLuhan on concrete poetry but—for the first time—to delineate concrete poetry's influence on those writings by McLuhan that are now foundational to media studies.

In the fourth chapter, "The Fascicle as Process and Product," I read digital poems into and out of Emily Dickinson's use of the fascicle. I assert the fascicle is a writing interface that is both process and product from a past that is becoming ever more distant the more enmeshed in the digital we become and the more the book becomes a fetishized object. Otherwise put, her fascicles, as much as the late twentieth-century digital computers and the mid-twentieth-century typewriters I discuss in chapters 2 and 3, are now slowly but surely revealing themselves as a kind of *interface* that defines the nature of reading as much as that of writing. More, extending certain tenets of media archaeology I touch on above, I read the digital into and out of Dickinson's fascicles as a way to enrich our understanding of her work. Such a reading is a self-conscious exploitation of the terminology and the theoretical framing of a present moment that is so steeped in the digital that, often without our knowing, it saturates our language and habits of thought.

Finally, in the postscript, "The Googlization of Literature," I focus on the interface of the search engine, particularly Google's, to describe a collection of literary contributions to contemporary media studies: works of *readingwriting* that explore a twenty-first-century media poetics as they question how search engines answer our questions, whether we ask them or not; how they read our writing; and even how they write for us. Readingwriting is literature like we've not seen it before. While building on

INTRODUCTION   xxi

a lineage of twentieth-century computer-generated texts, these works still give us a poetics perfectly appropriate for our current cultural moment in that they implicitly acknowledge we are living not just in an era of the search engine algorithm but in an era of what Siva Vaidhyanathan calls "The Googlization of Everything." But readingwriters who experiment with/on Google are not simply pointing to its ubiquity; they are also implicitly questioning how it works, how it generates the results it does, and so how it sells ourselves and our language back to us. These writers take us beyond the twentieth-century avant-garde's interest in the materiality of our own readingwriting to urge us instead to attend to the materiality of twenty-first-century *networked* digital language production. They ask, What happens when we appropriate the role of Google for our own purposes rather than Google's? What happens when we wrest Google from itself and instead use it not only to find out things about us as a culture but to find out what Google is finding out about us? "The Googlization of Literature," then, concludes *Reading Writing Interfaces* by providing an even more wide-ranging sense of a literary response to the interface-free.

Indistinguishable from Magic

Invisible Interfaces and Digital Literature as Demystifier

> The twenty-first century will not have the same craving for media. As a matter of course, they will be a part of everyday life, like the railways in the nineteenth century or the introduction of electricity into private households in the twentieth.
>
> —Siegfried Zielinski, *Deep Time of the Media: Toward an Archaeology of Hearing and Seeing by Technical Means*

Invisible, Imperceptible, Inoperable

If the twenty-first century does not have, as Siegfried Zielinski writes in the chapter epigraph, a craving for media, it is because media, by way of interface, are steadily making their way toward invisibility, imperceptibility, and inoperability. We cannot crave whatever is ubiquitous. As I describe in this section, contemporary claims about ubiquitous computing (ubicomp) as the definitive technological innovation of this century—supposedly, the third wave of computing, which replaces desktop computing and whose devices are seamlessly embedded throughout our everyday environment—consistently tout the invisibility of its interfaces as providing us with a more natural, more direct, inherently better way to interact with our computers and more generally with the world around us. Without attention to the ways in which interfaces are anything but invisible in how they frame what can and cannot be said, however, the contemporary computing industry will continue unchecked in its accelerating drive to achieve the perfect black box not only through the latest ubicomp devices but also through parallel developments, such as so-called Natural User Interfaces, Organic User Interfaces, and even the now widely prevalent multitouch interfaces. All of these interfaces share a common goal underlying their designs: to efface the interface altogether and so also efface our ability to read, let alone write, the interface, definitively turning us into consumers rather than producers of content. By contrast, with a critical eye on interface, a growing body of digital literature courts difficulty, defamiliarization, and glitch as antidotes to this receding present. Mark Weiser, the reputed father of ubicomp, originally believed that this mode of computing was an antidote to windows and desktop computing—now, we need digital literature as an antidote against what ubicomp has become.

Though this chapter focuses on invisible interfaces of the present and near future, as well as works of digital literature that disrupt this insistent drive toward invisibility, for the moment it is instructive to turn to the mid-1990s. This time period acts as a hinge that opens, in one direction, onto the subject of this chapter and, in the other, onto the subject of chapter 2, the turn from the 1970s' philosophy of open hardware/software to the mid-1980s' ideology of the user-friendly via closed hardware/software—a hinge that I hope demonstrates how we can wield media archaeology as a conceptual knife that cuts into the present and the near future, not just, in the sense of Zielinski's *deep time*, into the past, as in archaeology's digging in and around a historical context for a hole in the ground or the archaeological record. In 1995 Friedrich Kittler declared, "There Is No Software," as the logic of the computing community dictated that "in a perfect gradualism, DOS services would hide the BIOS, WordPerfect the operating system, and so on and so on on."[1] So while writer Rob Swigart noted in 1994 the gradual disappearance of the metaphorical desktop from his awareness—asserting, "That is the real danger. . . . Unless we pause from time to time to consider how these metaphors work to create boundaries . . . they will control us without our knowledge"—just a year later there would be no software at all.[2] Pivoting from the mid-1990s toward the present-future, not only does software obscure hardware, but interface obscures software.

We no longer have access to digital tools for making; instead, we have predetermined choices. Ideally, the seamlessness of ubiquitous computing devices will make even choice itself recede into the background. In this imagined near future, things will simply happen and we will simply do.

Thus, continuing in the direction of Kittler's 1995 essay, while Steven Johnson's 1997 *Interface Culture* was prescient in many different respects, one of his central claims was, "The most profound change will lie with our generic expectations about the interface itself. We will come to think of interface design as a kind of art form—perhaps the art form of the next century."[3] Although this declaration has held true in a certain respect, as evidenced by the digital writers I discuss in this chapter, our expectation that a user-friendly interface be an invisible interface has produced a present-future in which interface as an art form exists solely on the margins of digital literature and art as a means not to elevate the interface as a harmonious, beautiful objet d'art but by which to bring the interface back into view again via failure, discomfort, and dissonance. While Johnson did accurately foresee a future in which a "functional interface subculture" thrived, the conceptual framework underlying most definitions of *subculture* is one of oppositionality—no doubt drawn from a notion of the early twentieth-century avant-garde as that which pits itself against the mainstream, the ordinary, the status quo in favor of the marginal, the strange, the disruptive. This notion of the avant-garde as oppositional is not necessarily inaccurate, as Dada and Futurism did indeed see themselves as embattled movements that were explicitly against conventions and cultural norms of every kind. Just as certain Dada and Futurist practitioners worked from within language, painting, and music to undo linguistic, artistic, and musical conventions, so too certain digital writers and artists work to critique (by drawing attention to) the way in which not only hardware/software is now utterly black-boxed but its closed architecture is being marketed as a feature via attractive packaging that touts the marvelousness of natural, intuitive, invisible, and even "magical" interfaces. Ultimately, this literary critique seeks to undermine what is now an ideology of invisible interface design by disrupting from within the strictures of widely used interaction systems such as the webpage, broadly speaking, or, more specifically, the hyperlink. Now, digital interfaces are artful only to the extent that they don't work, which is now the only extent to which we can experience them at all.

Since the goal of having ubiquitous, invisible interfaces and digital devices has been achieved so definitively, the current model for interface subculture is not oppositional—for how can anyone oppose that which we cannot see, that which is as ever present as air—but rather *insurgent,* coming from within often via the efforts of both everyday users and more established digital writers and artists who creatively find ways to hack closed interfaces. In the following sections, I first trace several directions in contemporary interface design—working back from contemporary, slick ubicomp-related devices and interfaces to the now nearly pervasive multitouch interface. Then, I show how writers who work with and against the iPad (such as Jörg Piringer, Jason Edward Lewis, and Erik Loyer), who create codework (such as Mez [Breeze] and Nick Montfort), and finally, who create hypertext/Web-based work (such as Deena Larsen, William Gibson, Talan Memmott, Judd Morrissey, Jason Nelson, and Young-Hae Chang Heavy Industries) advance an insurgent twenty-first-century poetics by producing digital literature that is deliberately difficult to navigate or whose interfaces are anything but user-friendly.

Natural, Organic, Invisible

While this section is largely about ubicomp, in many ways ubicomp is a convenient stand-in for a wealth of contemporary interface designs, all stemming from interpretations, usually oversimplifications and misconstruals, of Mark Weiser's writings from 1988 to 1996 on what interface design could and should be. Weiser's ubicomp articles are surely responsible for introducing the term *invisible* into the lexicon of interface design, defining *invisibility* as a device's ability to be simultaneously everywhere yet also unexceptional in how it ideally lacks a distinct identity—the very opposite of the new highly visible, highly branded interface designs that claim a deep affiliation to ubicomp. By contrast, designers of the Fluid User Interface (Fluid UI), Organic User Interface (OUI), Natural User Interface (NUI), and even the first affordable multitouch interface demoed by Jeff Han, all consistently use *invisible* interchangeably with *natural* to describe their interfaces, so that both terms now imply a minimalist design, one that supposedly disappears and that is all the better equipped to mask the restrictiveness made possible by these interfaces that tightly control user access for the sake of becoming perfect portals for the consumption of content. These "invisible" and "natural" interfaces are also all marketed, of course, in the most joyful terms, to celebrate the fact that these devices sense *for us* what information we need and want.

From the MIT research group working on the Fluid UI, we are told their aim is to make "the user experience more seamless, natural and integrated in our physical lives" by creating interfaces that "perceive the user, her current context and actions and offer relevant services and information based on that awareness."[4] From the designers of the OUI, we read about a wondrous world populated by computers "with displays that are curved, flexible and that may even change their own shape in order to better fit the data, or user for that matter."[5] In OUI design computers are no longer distinguishable from the world in which they live, as their designers look toward "a final frontier in the design of computer interfaces that turn the natural world into software, and software into the natural world." This world of flexible surfaces is supposed to allow greater creativity, so that if you tire "of the color of your suit, the pattern of your wallpaper, or the interface on your cellphone, you simply download a new one from an online store," as if a world in which we choose from prefabricated surfaces and predetermined designs is the realization of creative living (see Figure 1).[6] From those working on the NUI, we find that it is an "interface that is effectively invisible, or becomes invisible to its user with successive learned interactions," and that *natural* is defined as "organic, unthinking, prompted by instinct."[7] Claims that ubicomp-related interfaces are more "natural" for "human beings" are echoed even by independent writers unaffiliated with any particular company or research group: "Human beings are physical creatures; we like to interact directly with objects. We're simply wired this way. Interactive gestures allow users to interact naturally with digital objects in a physical way, like we do with physical objects."[8] Finally, in a decisive attempt not to reframe the interface as even more invisible or more natural but rather to do away with it altogether, we read of predictions from IBM that within five years our brains will be synced with computing devices so that "if you just need to think about calling someone, it happens."[9]

Again, all of the foregoing interface designs imply a belief in the value of an interface that recedes from view, ideally to the point of invisibility, which now also implies inaccessibility. We need not know how it works, or how it works on us rather than us on it. As Adam Greenfield astutely pointed out in 2006, it's not only that these ubicomp-related devices make it possible for users to engage with them "inadvertently, unknowingly, or even unwillingly" but also that the discourse of invisibility, which he called the "discourse of seamlessness," "deprives the user of meaningful participation in the decisions that affect his or her experience."[10] Thankfully, in addition to Greenfield, a few critics, such as Ben Schneiderman, Catherine Plaisant, and Donald Norman, consistently point out that spatially or visually based interfaces are not necessarily improvements even over command-line interfaces, especially for those who are blind or vision impaired.[11] More, the supposed naturalness of ubicomp-related gestural interfaces is utterly misleading once we consider that "most gestures are neither natural nor easy

WHAT MAKES AN INTERFACE FEEL ORGANIC?

BY CARSTEN SCHWESIG

M ovies can make us forget that we are sitting in a cinema among strangers, looking at images projected onto a wall. Instead, we feel as though we are observing real people in real situations and we become emotionally involved in the narrative. User interfaces can trigger a similar suspension of disbelief: we forget we are operating a machine to manipulate virtual, digital data. Instead, we experience media and applications as part of our physical environment. Such interfaces feel "natural," or rather "organic."

Figure 1. Gummi interface prototype showing map navigation.

FIGURE 1. *From a special issue of* Communications of the ACM *on Organic User Interfaces (OUIs), we are told that user interfaces such as the OUI can trigger the same suspension of disbelief as when we go to the movies—because of how "natural" or "organic" they feel, both movies and OUIs are, it's implied, magical.*

to learn or remember. Few are innate or readily predisposed to rapid and easy learning. Even the simple head-shake is puzzling when cultures intermix. . . . Similarly, hand-waving gestures of hello, goodbye, and 'come here' are performed differently in different cultures."[12]

Even more surprising than the unthoughtful claims about seamlessness, invisibility, and the nature of human beings are the techno-determinist assumptions about how ubicomp-related devices *will* be deployed everywhere in the future and how this imagined deployment necessarily implies "the inadequacy of the traditional user interface modalities we've been able to call on, most particularly keyboards and keypads."[13] Again, marketing rhetoric convinces us that these interfaces work more "naturally" than what one designer calls the "crap desktop," which another claims is simply an outdated mode of interaction that "severely constrains us."[14] The rhetoric might not be so disagreeable if it didn't also help determine the shape of the future of computing—one that, for these designers, would ideally be populated not even with computers as appliances but with appliances embedded within small computers.

It's worth underscoring that the rhetoric around ubicomp is indeed just that, for most of its devices have turned out to be resounding failures. Whereas Mark Weiser advocated for what he believed was a better way for us to interact with computers—one with computers so small, so plentiful, so uniquely tailored to specific tasks, and so unimportant that human-to-human interactions would become dominant over individual interactions with branded personal computers made for multitasking—companies like Samsung have no such ethical investment in their Wi-Fi-enabled refrigerator "pre-loaded with apps," of course made only for Samsung, that allow you to check Twitter, look up recipes, or listen to Pandora. It turns out the future is not one in which, as Weiser heralded in 1996, we "most fully command technology without being dominated by it."[15] Instead, the future of computing is domineering, branded, and boring.

We can see a clear arc in Weiser's writing on ubicomp from this point in 1996 back to when he first coined the term in 1988 while serving as head of the Computer Science Laboratory at Xerox PARC. Cowritten with Jeff Sealey, his "The Coming Age of Calm Technology," which I quote in the preceding paragraph, signaled his concern that the philosophy driving most computing devices was one grounded in a paternalistic notion of ubiquity through invisibility that took the form of inaccessibility rather than a ubiquity of "calm technology," technology that "engages both the center and the periphery of our attention, and in fact moves back and forth between the two." He wrote, "Designs that *encalm and inform* meet two human needs not usually met together."[16] Illustrating just one of many reversals over the course of the history of computing, the goal of ubiquitous computing was never, as it is now, to transform the value of invisibility into an elimination of the need to freely access tools and information or the need to understand computer processes altogether. Simply because something has the ability to move to the periphery of our attention does not preclude us being aware of it or understanding how it works.

Just a few years earlier in his 1994 "The World Is Not a Desktop," Weiser even advocated for humanists to understand invisibility as "they specialize in exposing the otherwise invisible."[17] More, while he recognized in this same article that "a good tool is an invisible tool," writing, "I mean that the tool does not intrude on your consciousness; you focus on the task, not the tool," he did not believe that invisibility in computing should mean making computers appear more human-like at the cost of accessing the underlying computer:

Why should a computer be anything like a human being? Are airplanes like birds, typewriters like pens, alphabets like mouths, cars like horses? Are human interactions so free of trouble, misunderstanding, and ambiguity that they represent a desirable computer interface goal?[18]

Therefore, neither did he advocate using "magic" as a way to trick the user into thinking the computer was behaving like a human by doing something it was not, usually via attractive packaging that called attention to the computer even more:

> Take magic. The idea, as near as I can tell, is to grant wishes. . . . I wish my computer would only show me what I am interested in. But magic is about psychology and salesmanship, and I believe a dangerous model for good design and productive technology. The proof is in the details— magic ignores them. Furthermore magic continues to glorify itself, as Robin Williams' attention-grabbing genie in Aladdin amply illustrates.[19]

But moving back in time, when ubicomp came to the attention of the general public in 1991 via a *Scientific American* article provocatively titled "The Computer for the 21st Century," Weiser framed ubicomp not with the twin terms *encalm* and *inform* but rather with the value-laden term *invisibility,* which has continued to dominate the rhetoric around nearly every new computing interface that has emerged since then. When Weiser first introduced ubicomp to the general public, he opened with the declaration, "The most profound technologies are those that disappear. They weave themselves into the fabric of everyday life until they are indistinguishable from it." Weiser went on to cite print as a literary technology that had achieved this level of usability, familiarity, and thus invisibility.[20] Although this example of print as a technology that "gets out of the way of the user" has been used repeatedly in subsequent years to explain how ubicomp devices give us the opportunity to no longer have to "continuously tinker with the system, maintaining it and configuring it to complete a task," Weiser's original use of print as an example of effective ubicomp was meant in an entirely different spirit—it was, instead, about widespread availability, portability, convenience, flexibility, and readily transmitted information via ubicomp devices called *tabs, pads,* and *boards.*[21]

Quite unlike any contemporary ubicomp or ubicomp-related device, Weiser's tabs, pads, and boards were all developed at Xerox PARC to allow the user to customize what and how much information was displayed. Pads, for example, were supposed to be something between a sheet of paper and a laptop computer. Despite the family resemblance, these pads were profoundly different from the twenty-first-century iPad. As he wrote, "The pad that must be carried from place to place is a failure. Pads are intended to be 'scrap computers' (analogous to scrap paper) that can be grabbed and used anywhere; *they have no individualized identity or importance.*"[22] Moreover, diametrically opposed to the iPad, which in many ways represents the logical endpoint of windows, Weiser's pads were "an antidote to windows. Windows were invented at PARC and popularized by Apple . . . as a way of fitting several different activities onto the small space of a computer screen at the same time. . . . Pads, in contrast, use a real desk. Spread many electronic pads around on the desk, just as you spread out papers. Have many tasks in front of you."[23] Finally, the picture of the pad in Figure 2, displaying its inner structure, hearkens back to another, earlier era of computing which valued an open (and therefore extensible) architecture.

Nowadays, introducing the latest iPad to the general public by opening up the device is practically unthinkable. In fact, at the Apple event to unveil the fourth-generation iPad and the iPad mini in October 2012, when Phil Schiller said, "Let's open it up and see what's inside," "inside" amounted to a screenshot of Apple-branded icons representing different functions and components. It's also not surprising that Weiser's later arguments against computing modeled on magic, via what he called "psychology and salesmanship," have been recently reversed and used even as a selling point for interfaces that do anything but encalm and inform. The iPhone/iPad multitouch interface, which is constantly touted as "magical" or as something that allows us to perform "magic tricks," is invisible in the sense that it constantly seeks to hide its inner workings through glossy, attractive packaging that makes the iPhone/iPad highly visible

KEY COMPONENTS OF UBIQUITOUS COMPUTING are the pads and tabs under development at the Xerox Palo Alto Research Center. The page-size pad (*top, exterior and interior views*) contains two microprocessors, four million bytes of random-access memory, a high-speed radio link, a high-resolution pen interface and a black-and-white display that is 1,024 by 768 pixels. Because it uses standard window system software, the pad can communicate with most workstations. The much smaller tab (*at left*), $2^{3}/_{4}$ by $3^{1}/_{4}$ inches, has three control buttons, a pen interface, audio and an infrared link for communicating throughout a room. The author believes future homes and offices will contain hundreds of these tiny computers.

very small, and the range large (50 to 100 meters), so that the total number of mobile devices is severely limited. The ability of such a system to support hundreds of machines in every room is out of the question. Single-room networks based on infrared or newer electromagnetic technologies have enough channel capacity for ubiquitous computers, but they can work only indoors.

Present technologies would require a mobile device to have three different network connections: tiny-range wireless, long-range wireless and very high speed wired. A single kind of network connection that can somehow serve all three functions has yet to be invented.

Neither an explication of the principles of ubiquitous computing nor a list of the technologies involved really gives a sense of what it would be like to live in a world full of invisible widgets. Extrapolating from today's rudimentary fragments of embodied virtuality is like trying to predict the publication of *Finnegans Wake* shortly after having inscribed the first clay tablets. Nevertheless, the effort is probably worthwhile:

Sal awakens; she smells coffee. A few

minutes ago her alarm clock, alerted by her restless rolling before waking, had quietly asked, "Coffee?" and she had mumbled, "Yes." "Yes" and "no" are the only words it knows.

Sal looks out her windows at her neighborhood. Sunlight and a fence are visible through one, and through others she sees electronic trails that have been kept for her of neighbors coming and going during the early morning. Privacy conventions and practical data rates prevent displaying video footage, but time markers and electronic tracks on the neighborhood map let Sal feel cozy in her street.

Glancing at the windows to her kids' rooms, she can see that they got up 15 and 20 minutes ago and are already in the kitchen. Noticing that she is up, they start making more noise.

At breakfast Sal reads the news. She still prefers the paper form, as do most people. She spots an interesting quote from a columnist in the business section. She wipes her pen over the newspaper's name, date, section and page number and then circles the quote. The pen sends a message to the paper, which transmits the quote to her office.

Electronic mail arrives from the company that made her garage door opener. She had lost the instruction manual and asked them for help. They have sent her a new manual and also something unexpected—a way to find the old one. According to the note, she can press a code into the opener and the

missing manual will find itself. In the garage, she tracks a beeping noise to where the oil-stained manual had fallen behind some boxes. Sure enough, there is the tiny tab the manufacturer had affixed in the cover to try to avoid E-mail requests like her own.

On the way to work Sal glances in the foreview mirror to check the traffic. She spots a slowdown ahead and also notices on a side street the telltale green in the foreview of a food shop, and a new one at that. She decides to take the next exit and get a cup of coffee while avoiding the jam.

Once Sal arrives at work, the foreview helps her find a parking spot quickly. As she walks into the building, the machines in her office prepare to log her in but do not complete the sequence until she actually enters her office. On her way, she stops by the offices of four or five colleagues to exchange greetings and news.

Sal glances out her windows: a gray day in Silicon Valley, 75 percent humidity and 40 percent chance of afternoon showers; meanwhile it has been a quiet morning at the East Coast office. Usually the activity indicator shows at least one spontaneous, urgent meeting by now. She chooses not to shift the window on the home office back three hours—too much chance of being caught by surprise. But she knows others who do, usually people who never get a call from the East but just want to feel involved. The telltale by the door that Sal pro-

FIGURE 2. *Images of Mark Weiser's ubicomp pad and tab as they appeared in* Scientific American *in 1991.*

and puts it at the center of our attention while becoming a fetishistic object that's anything but Weiser's scrap pads.[24]

The iPad: "A Truly Magical and Revolutionary Product"

On October 23, 2012, Apple's Tim Cook and Phil Schiller "unveiled" (the word of choice to describe every introduction of a new computing device, a word that evokes a magician revealing a trick's hidden mechanism) the new iPad mini, along with several other products that were updated with sharper displays or to be thinner, faster, smaller. Given the mini's dimensions, 7.87-by-5.3 inches and 0.28 inches thick (what literature about the mini doesn't make its dimensions of utmost importance?), Cook and Schiller mentioned "thin," "thinner," or "thinnest" throughout the one-hour-twelve-minute event no less than thirty-five times. "Incredible" or "incredibly" were a close second at twenty-seven times, and "amazing" was third, at twenty-two times—and as if lifted from a women's fashion magazine, "beautiful," "elegant," "gorgeous," and "light" were consistently peppered throughout.[25] The mini is "a quarter thinner than the fourth-generation iPad. To put it in context, it's as thin as a pencil. It weighs just 0.68 pounds. That's 53 percent lighter than the fourth-generation iPad. . . . It's as light as a pad of paper."[26] In the usual breathless tones of an Apple "event," as they're called (again, no other term better describes the quiet theatricality of these product launches), Schiller was careful to repeat that the device was not "just a shrunken down iPad," and again, on the Apple website we are reminded the iPad mini "isn't just a scaled-down iPad. We designed it to be a concentration, rather than a reduction, of the original."[27] No unveiling would be complete without plenty of discussion about the fine craftsmanship that went into its construction, as "every detail is finely crafted and made exquisite," coupled with declarations about how "it's beautiful on both sides" and also "beautiful from every angle" because of its aluminum and glass enclosure, instead of the "heavy plastic" used by other products. Apple's quest for

thinness, particularly through its line of iPads—whose gestural, multitouch interface is a direct descendent of Weiser's
ubicomp—is a quest for an immensely powerful device that
moves as close as possible to invisibility without disappearing
altogether, for then we couldn't marvel at its highly branded,
highly individual, and supposedly artful packaging. This quest
for the paradoxical combination of beauty, thinness, and invisibility most obviously extends back to the release of the iPod
in October 2001. As Steven Levy writes in his usual awestruck
tone, with an appearance by Jony Ive midsentence, "The iPod
was the boldest step yet toward whiteness, an effort directed
to the heart of visual simplicity and minimalism, with perhaps
a yearning toward invisibility. 'Right from the very first time,
we were thinking about the product, we'd seen this as stainless
steel and white,' Ive explained. 'It is just so . . . so brutally simple. It's not just a color. *Supposedly neutral—but just an unmistak-
able, shocking neutral.*' "[28]

This is where magic comes in—through the supposedly neutral. The iPad's packaging, part of which is dubbed an "enclosure," no doubt for the word's undercurrent of specialness or
even awe, and the device's marketing rhetoric are so seductive
that we consistently overlook the fact that we are willingly
suspending disbelief every time we use it. In fact, the willing
suspension of disbelief is a key component to magic shows, for
although the audience wants to be amazed by feats that are
seemingly impossible, their amazement depends on two key,
interdependent components: they must believe that the magician's assistant is not actually being sawed in half or that a
dove is not actually being turned into a handkerchief, and yet
they must remain in the dark (literally and figuratively) about
exactly how the trick works.[29]

This logic was most clearly at work during the January 2010
launch of the first iPad, at which Steve Jobs stood onstage in a
dimly lit auditorium (that itself looked like a modernized early
twentieth-century theater, with its ornate friezes and columns
juxtaposed with the clean lines of a black stage) and opened

the launch by calling the iPad "a truly magical and revolutionary device" before pulling the device out from underneath a
black cloth on a pedestal.[30] Over the next one hundred minutes, Jobs went on to celebrate this device that was so "gorgeous," "incredible," "extraordinary," "awesome," "amazing,"
"phenomenal," and "unbelievable" and that was also their
"most advanced technology in a magical and revolutionary
device at an unbelievable price." Most telling, throughout the
show a range of Apple executives explained that using the iPad
was "just that simple" (repeated at least ten times) because "it
just all works." "You don't even think about it. You just do."
This reminder that the iPad transcended thought was only
the most recent and most obvious example of similar Apple
product slogans. "It just all works" was a near perfect echo of
Apple's 2007 ad for Mac OS X, which also "just works."[31] If after
ninety minutes of this show you were not quite convinced of
the iPad's bewitching properties, Jony Ive, Apple's senior vice
president of industrial design, appeared on screen to tell you:
"It's true—when something exceeds your ability to understand how it works, it sort of becomes magical. That's exactly
what the iPad is."[32] Ive was clearly channeling science fiction
writer Arthur C. Clarke's famous Third Law, which states, "Any
sufficiently advanced technology is indistinguishable from
magic."[33] The difference here is that the iPad, which is indeed
an advanced piece of technology, was deliberately made to appear magical. It's not that one day we will look back and, with
clear hindsight made possible by a more refined understanding, comprehend the iPad and no longer see it as magical. Ideally, we will never comprehend it. The iPad works because users
can't know how it works.[34]

    With wild techno-enthusiasm, Jesus Diaz's writing for *Giz-
modo* on April 2, 2010, the day before the release of the iPad,
perfectly represents the irresistible pull of these new, slick computing devices. Without even a hint of critical-mindedness, he
regurgitates some of the same language used to sell the iPad
four months earlier at the product launch:

[The iPad] shows that *computers have—must—be an invisible platform,* one that shifts its appearance to give people the tools to complete the tasks they want to accomplish, whatever these are. . . . By being invisible and letting the applications do the work in the most simple way possible, the power of the computer will, at last, be available for everyone. No previous knowledge required. From a 3-year-old baby to your 90-year-old grandma, *people will be able to just do things.*[35]

Diaz's rhetoric is, on the surface, remarkably similar to Mark Weiser's on ubicomp, but the fundamental difference is that Diaz's notion of an invisible computer whose appearance is constantly shifting and that "just does" depends on something that might as well be called "magic," which is, again, precisely what Weiser argued against.

Not surprisingly, the iPad launch was followed by an ad campaign throughout 2010 that included commercials such as "iPad Is Magical"—which doesn't mention "magic" once and instead gently nudges viewers into thinking the device must be magical since it "is" (rather than enables or gives tools for) "medical, live, musical, work, play, memories, social, magazines, historic."[36] This commercial was followed a few months later by "iPad Is Delicious," which claims the device is so because it is "current, learning, playful, literary, artful, friendly, productive, scientific, magical" (of course, "literary" and "artful" are illustrated with users merely reading, flipping pages, and finger-painting).[37] Then, a few months later, Apple released "iPad Is Electric," this time because it is "cinematic, elementary, academic, full size, presenting, bought, sold, fantasy."[38] Perhaps, then, it is perfectly fitting that the iPad (or perhaps just its marketing campaign) has given rise to so-called iPad magicians, who capitalize on the way in which users/consumers so easily and willingly suspend disbelief. Employing the device as a twenty-first-century version of a black hat, Charlie Caper and

but an animal might work.

FIGURE 3. *Shinya the "Salary Magician" stands outside an Apple store in Japan and creates the illusion of producing a dove from his iPad.*

Erik Rosales use the iPad as a magic prop to convince spectators to invest in Stockholm real estate. Shinya the "Salary Magician" turns the image of a dove on his iPad into an actual dove that flies in front of an Apple store in Japan (see Figure 3). And Simon Pierro, the "Wizard of OS," pours beer from the device in a German tavern to the awe and delight of a nonstop stream of patrons (see Figure 4).[39]

The iPad is, without a doubt, the most influential, "magical and revolutionary" closed computing device of the twenty-first century—and for the skeptical, Apple has the sales numbers to prove it. Tim Cook proudly intoned in a friendly southern drawl at the iPad mini launch that Apple had sold 100 million iPads in the past 2.5 years; that in November 2012 alone they had sold three million devices in just three days; that iPads accounted for 91 percent of the total Web traffic; that 94 percent of Fortune 500 companies were testing or deploying iPads; that of the

FIGURE 4. *Simon Pierro, the "Wizard of OS," creates the illusion of pouring beer from an iPad in front of delighted patrons in a German tavern.*

total 700,000 apps available through iTunes, 275,000 of those were specifically for iPad.[40]

Yet at the same time as iPad sales increase and the device becomes practically de rigueur in middle-class homes, workplaces, and schools, Apple continues not only to co-opt the terms *invisibility* and *user-friendly* but also—as I briefly point out in relation to the iPad's "literary" and "artistic" capabilities—to redefine the very notion of creativity, as if it has all along been undeterred in principle from its mid-1990s ad campaign to "think different." As Jobs said to his audience at Macworld in 1997, "You still have to think differently to buy an Apple computer. . . . The people who buy them do think different. They are the creative spirits in this world, and they're out to change the world. *We* make tools for those kinds of people."[41] Regardless of how much today's consumers of Apple products "think different," thinking can't overcome the brute fact that from Apple's perspective creativity on the iPad largely does not mean creating or producing content—neatly ensured by both its slick external packaging and, as I lay out in the proceeding section, its operating system.

If the iPad signals the future of computing and of ubicomp-related computers, then perhaps it also simultaneously signals a future generation of hackers who will be driven to find a way out of this flat notion of creativity that amounts to little more than consumption and manipulation as users are turned into audience members watching their devices perform magic tricks before their very eyes. (Incidentally, this notion of creativity couldn't be more at odds with the tinkerer/homebrew notion of creativity underlying the 1980 ad campaign for the Apple II, which invited users to write directly to the company and describe "the most original use of an Apple since Adam.") While there will always be users who find ways to produce content on any device—in fact, I touch on several innovative digital literature iPad apps—given the months or even years it might take a novice to learn the Objective-C programming language, which is the standard language for iOS development, not to mention the rigid and restrictive iOS guidelines, it remains that the iPad, both inside and out, is unquestionably made for consumption, and its wild popularity, driven or bound to Apple's marketing rhetoric, continues to determine the shape of computing as companies clamor for a share of the profits.

From Videoplace to iOS: A Brief History of Creativity through Multitouch

It is as if Apple has successfully turned *creativity* into a proprietary eponym like Kleenex or Frisbee. But against forgetting what creativity via computers *could* mean and in fact at one point *did* mean, creativity (not unlike invisibility), especially via multitouch devices, has undergone significant reversals over the past twenty to thirty years. Myron Krueger's Videoplace is a particularly poignant example of how *creativity* in computing at one time implied tinkering, active learning, and interactivity, rather than being a term leveraged to drive profit and that often means manipulating content by making surface-level changes—flipping through preprogrammed, locked-in settings and functions.

Krueger's work with artificial reality (AR), which he defined as the creation of synthetic, alternative realities, particularly through his Videoplace interface, is frequently cited as a crucial yet frequently overlooked influence on the development of multitouch. Of course, the history of multitouch interfaces is long and varied (himself a pioneer in multitouch interface design, Bill Buxton points out that keyboards were the first multitouch interfaces). But Krueger's work is essential not only because he was the first to create a wide and workable repertoire of gestures (including many gestures we now take for granted, such as the pinch and the swipe) that did not require gloves, headsets, mice, styluses, etc. but also because he was firmly invested in developing ways to interact with computers for aesthetic, scientific, and practical ends.[42] As Krueger puts it in a video from 1989 overviewing his work in Videoplace and responsive environments, he started work on these artificial reality systems after spending time teaching students "the essence of computers and trying to make it so that you would experience a computer . . . rather than doing something efficiently. And that is essentially the role of the artists—and I was thinking of expressing the computer the same way."[43]

In the opening to *Artificial Reality,* Krueger's account of his work in AR throughout the 1970s and 1980s, he makes the point that all of his work in interface design was grounded in his education at Dartmouth, whose attitude in the 1960s and early 1970s "was that knowledge of computers was part of a liberal arts education, and that anything we might do with these machines was likely to be instructive."[44] The point was to help students feel empowered to understand and create with computers—the diametric opposite of taking on an awestruck, passive stance. Thus, in 1972 Krueger began work on Videotouch, or what he called a "two-way installation," which encouraged two participants—each in separate virtual environments—to touch each other's projected screen image and thereby create a shared environment called Videoplace. Over time this interaction system came to include such a remarkably rich collection of gestures and multifinger, multihand, and multiperson interaction that by comparison contemporary devices such as the iPad seem like nothing more than pale imitations.[45]

Just as important as the gestures and forms of touch-based interaction he developed was that the project was a digital staging of defamiliarization that encouraged a processual, open-ended exploration of the unexpected. As Krueger puts it, "This new graphic experience can highlight assumptions and expectations of which we are never aware, because it does not occur to us that our world could be other than it is."[46] Should we make the mistake of thinking Krueger sees art or aesthetics as Apple does—as the passive enjoyment of the beautiful that happens through magical devices on which you "just do"—we learn that "the purpose of the [Videoplace] displays is to provide a context within which the interaction occurs. . . . This context is an artificial reality in which the laws of cause and effect *are composed by the artist. The beauty of the displays is not as important in this medium.*"[47] Rather than trying to efface the medium altogether to the point of near invisibility, what is important in this medium is the medium itself—that is, the goal of one's interactions and creations, the two inextricably intertwined, is "to express the medium itself."[48]

More, if the emphasis is on experiencing and expressing the medium itself through unexpected interactions, artistic production shifts away from representing things as they are to something more aligned with conceptual art and happenings from the 1950s and 1960s. This shift in turn works against the hermeneutic tradition, for rather than peel away layers of meanings to arrive at an interpretation, critics have very little choice but to simply describe the unfolding experience. One particular form of artistic interaction on Videoplace is called "body surfacing," which makes possible the continuous painting of the participant's image as she or he moves across a screen, while another is called "videosyncrasy," in which a participant uses his or her finger to trace a path that is then traveled by accelerating and decelerating pulses of light that have what Krueger calls "a decaying tail."[49] The interactions on Videoplace are, then, ones that are open, active, generative, and—given that the emphasis is on the processual nature of interactions and not their product—art in themselves. In terms of literary arts such as poetry, which is conventionally understood to concern itself with the expressive delivery of some particular insight that readers then interpret, once our attention is turned to the poetic process itself the result is an emphasis on the letters and the words themselves and the participant's (now reader's) interactions:

> The reader would enter into a relationship with the words, which would become entities moving about the screen, each with its own rules of behavior. There rules would be based on the aesthetic of the poet and on the words themselves. The intent of such an interaction would be to create a poetic experience, rather than to duplicate exactly the function of poetry.[50]

Note how every user of the system is also a poet, for simply by virtue of interacting, one creates. More, Krueger's emphasis on processuality and on expressing the specificity of the medium moves us toward the practice of poetics—the doing of poetry through an attention to the material dimensions of the letter, morpheme, or word. Continuing on, as if writing about the Marshall McLuhan–inspired concrete poetry from the 1960s and 1970s that I discuss in chapter 3, Krueger proclaims that "allowing a word to interact physically with a participant is a symbolic statement, for the written word is then no longer solely a vehicle for communicating meaning, but rather is an entity behaving on its own as well."[51] Through Videoplace one can explore and express it *as* a medium at the same time as one can explore and express the written word as another kind of medium. Videoplace is a medium for media study.

FIGURE 5. *Screenshot from a 1989 video of Myron Krueger demoing Videoplace and Videotouch.*

iPoems

Given the diametrically opposed pulls between the philosophy underlying Videoplace and that underlying contemporary multitouch devices such as the iPad, it's surprising that Krueger's description of what a truly interactive poetry could be in an AR environment sounds remarkably similar to what contemporary digital writers/artists have accomplished twenty years later with their iPad/iPhone apps. In digital poetry apps by Jörg Piringer, Jason Edward Lewis, and Erik Loyer, the acts of touching, tapping, swiping, and sliding are the only mechanisms by which the works come into being, and in some cases it is the only way by which the works unfold. That said, since these works are primarily about the medium of the iPad and the word as media, they are also not particularly invested in providing the reader/user with a text that is interpretable. If Lewis, for example, did not provide a description about what each of his poems were about, it would be difficult to say what the turning, twisting, moving letters and occasional phrases or sentences "meant" in relation to the work as a whole.

Even more striking is that these digital writers have produced innovative works that express the multitouch medium as a medium despite the Apple iOS developer guidelines that continue the interface design tradition of making sweeping generalizations about "people," assumptions about nature and intuition, all of which are tied to statements about the necessity of hiding the device's workings via interface and therefore the necessity to black-box the device. In the iOS Human Interface Guidelines, we read that "a great user interface follows human interface design principles that are based on the way people—users—think and work, not on the capabilities of the device"; that the multitouch display "encourages people to forget about the device and to focus on their content or task"; and that "iOS-based devices and the built-in apps are intuitive and easy to use, so people don't need onscreen help content."[52] There are also the many and varied difficulties associated merely with getting one's app approved—for one may create only within Apple's rigid and deeply moral strictures. For example, Apple is clear they will reject apps with easter eggs or "undocumented or hidden features inconsistent with the description of the app"; apps "that encourage excessive consumption of alcohol or illegal substances"; apps that "present excessively objectionable or crude content"; apps that include pornography, as well as gambling; and apps that "target a specific race, culture, a real government or corporation, or any other real entity."[53] First, many banned features, such as easter eggs, implicitly ban the unexpected—even though an experience of the unexpected was precisely what Krueger originally sought to foment through his multitouch display and even though the experience of the unexpected defines many compelling works of art and literature.[54] In fact, Apple maintains as much control as possible over apps by intentionally avoiding stating their criteria for determining what is objectionable, crude, porn, targeting, etc. Neither are they clear about how app reviewers discern between encouragement and mere or, even, pointed representation.

Despite this remarkable range of restrictions on both the form and the content of creative expression on the iPad, in a handful of digital literature apps there is still a clear connection between Videoplace and the attempt to explore and express the medium as a medium through open-ended play, touch, embodied movement, and a courting of the unfamiliar (*through* the unfamiliar). For example, Piringer's *abcdefghijklmnopqrstuvwxyz* is essentially a playful DIY kinetic poetry platform that allows users to flick any or all letters of the alphabet onto a simulated white canvas (see Figure 6). While the user can, to some degree, control the movement of the letters by tilting and rotating the iPad (taking advantage of the accelerometer, a sensor-based technology capable of measuring the force of gravity- or movement-induced acceleration), the user can also choose four modes in which the letters can move without requiring any additional interaction.[55] "Gravity" causes letters to drop and bounce while emitting an oral articulation of the letter-form every time it bounces off one of four virtual surfaces. "Crickets" causes letters to turn into pixilated critters, emitting an equally pixilated sound and traveling a more indeterminate path on a plane that's horizontal rather than the vertical plane of "gravity." "Vehicles" and "birds" also operate on this horizontal plane. Taking advantage of the blank, white canvas that Piringer can use to suggest nearly any surface, "vehicles" not surprisingly causes letters to behave and sound like automobiles moving across the ground, whereas we are invited to imagine the letters in "birds" moving across the sky and at a much slower speed. (The app makes just as much room for destruction as it does for creation, as it includes the capability of targeted destruction of individual letters on-screen or detonation of the entire alphabetic scene.) Here, individual letters become entities unto themselves, and our experience of them is one that defies the standard procedures for literary analysis.

Jason Lewis's works from his P.O.e.M.M series (Poetry for

FIGURE 6. *Screenshot from Jörg Piringer's* abcdefghijklmnopqrstuvwxyz *app using all four modes to mobilize letters: "gravity," "crickets," "vehicles," and "birds."*

Excitable [Mobile] Media) are also relevant examples of apps that do the work of inventively expressing the iPad's unique multitouch capabilities. *What They Speak, Migration,* and his most recent limited-edition app, *Smooth Second Bastard,* all—regardless of the author notes accompanying each work that state what the poetry app is about—embrace an aesthetic of exploring, only through touch, the material, tangible, yet ephemeral qualities of individual letters and words. *What They Speak* is the first in the P.O.e.M.M series, and it allows the user to draw tracks of text (either from the ready-made letters and words, from a poem the user writes, or from text drawn from Twitter) that read backwards, with a swipe to the right, and forwards, with a swipe to the left.[56] *Migration* is perhaps even more mysterious, as it features vague spermatozoa-like entities that similarly respond to tapping and swiping and have short phrases trailing out of them (such as, "I'm not sure if this is happenstance").[57] Finally, Lewis's most recent app, *Smooth Second Bastard,* seems to be an experiment to see how far the iPad multitouch interface can be made into a complex and generative interface for the experience of a kind of procedurally based poetry (see Figure 7).[58] This limited-edition app (in itself an oddity, but one that reminds us that we never own the apps we purchase; we are only granted access to them, if not by the app creator then by Apple) utilizes touch first as a way to generate spools of text from either side of the user's pressed finger. Without the pressure of the finger, all but one word disappears, and as Lewis explains, "After three words have built up, each new word—created by releasing a line—leaves behind one letter as the rest disappears off-screen. The lines, the words, and the letters all form their own texts, creating a three-dimensional poem."[59] If the poem is about anything at all, it is about experiencing the complexity of its touch-driven, generative medium through the additional medium of language itself.

Finally, Erik Loyer's *Strange Rain* particularly resonates with Krueger's vision for creativity via Videoplace (see Figure 8).[60] The app contains three different modes of falling rain and/or

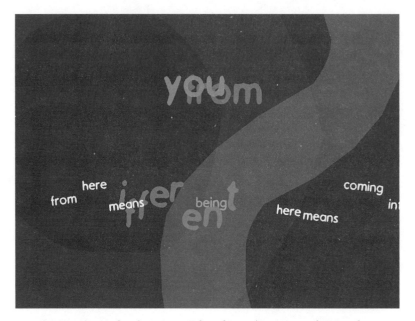

FIGURE 7. *Screenshot from Jason Edward Lewis's app* Smooth Second Bastard.

text that respond to tilting, rotating, and touching and that insist on sustained interaction by touch—not by scrolling, flipping, clicking, or viewing—as a way to immerse oneself deeper in the mechanics of the app. The first mode is the slow, meditative "wordless" mode, which turns the iPad into a window onto which rain falls down (or depending on how you wish to hold it, the rain can even fall up), settling in splattered patterns on the screen. Swiping or swirling creates a responsive pattern of fallen raindrops, and pinching in or out changes the intensity of the rain, which as the author quietly notes at the bottom of the screen, can also "perform" as an element of the equally strange soundtrack playing at the same time that has neither a beginning nor an end. As another hint at the bottom of the screen tells us, "The more times you play through the melody, the more strange things will appear." For example, planes ominously float across the stormy sky; frames disappear into frames of frames

FIGURE 8. *Screenshot from the "wordless" mode in Erik Loyer's app*
Strange Rain.

of the rain-covered window (making us suddenly aware of the
window and even the device itself as media that frame our experience); and the white and grey of the scene abruptly change
to red and green.[61] As Loyer himself puts it, "Before your eyes
and beneath your fingers, the familiar becomes strange, and

the strange, familiar." The "whispers" mode builds on the structure of the rain-spattered window and adds a feature by which some raindrops turn into words such as "absolve," "liberate," or "nourish."[62] "Story" mode imports a story Loyer wrote, "Convertible," into the multitouch environment as a way to explore the impact of the same interactions with rain on story—that is, in this mode touching the screen produces text expressing thoughts from the main character, Alphonse, who has stepped into his sister's rainy backyard to clear his head after what we can only assume was an earlier conflict inside the house. The iPad screen then turns into the eyes of a character looking up into rain falling from the sky—prompting us to think of eyes as media just as much as windows or screens—and tapping makes brief thought statements appear, whereas dragging produces further elaborations on these same thoughts. *Strange Rain* shows us that—quite in spite of Apple—it is possible to create apps that help us think through and experience the multitouch device as both interface and medium.

Making the Invisible Visible: Hacking, Glitch, Defamiliarization in Digital Literature

It may be precisely because our devices are ever more hermetically sealed that *hacking* is an apt term to describe certain works of digital literature created before 2000, before the era of the magical device. Although this section is mostly concerned with glitch, understood as intentional disruptions to the smooth surface of the interface, I touch on a long-standing tradition in innovative writing that helped make way for these glitch works. This tradition took a hacker's approach to both writing and media-specific interface, often doing so by drawing attention to the process underlying the writing product, the way in which process and product were unavoidably intertwined. These works engaged in hacking not in the more recent sense of illegally bypassing computer security mechanisms but rather in its earlier (perhaps original) sense, embodied by the computer hobbyists of the Homebrew Computer Club from the 1970s and early 1980s, who were invested in the communal enterprise of open-source DIY computing. Hacking in this sense has been usefully reenlivened by McKenzie Wark, who describes it in terms of the activities of a class of people who "create the possibility of new things entering the world" and whose slogan is, "Not the workers of the world united, but the workings of the world untied."[63]

Both early and contemporary examples of codework digital literature untie the workings of the computer not just by making visible the code or the normally invisible underbelly of our digital devices but by making the code the work of literature itself. Process becomes both product and fodder for appropriation and remix by others. Although not particularly invested in glitch, difficulty, or failure, the Apple BASIC code poem buried

```
1.SocialConnectionAccessProtocol[- SCAP -]

SocialConnectionAccessProtocol[- SCAP -]
ControlVersioningSystem
09:07am 25/05/2007


% cvs -d :codependentserver:internaltripwiring@cvs-mirror.abortive.org:/cvsroot login
(Logging in to internaltripwiring@cvs-mirror.abortive.org)
CVS password: [internaltripwiring]
%
% cvs -z3 -d :codependentserver:internaltripwiring@cvs-mirror.abortive.org:/cvsroot co SCAP

cvs server: Updating abortive/directory/SCAP
Ur abortive/directory/SCAP/NO.pls
Ur abortive/directory/SCAP/YES.dmg
Ur abortive/directory/SCAP/ChangeRealityLog


--
#There are currently 6 SocialConnection release_valves available.
#Use the main an[ti]xiety_loading trunk:
#Release-1_0: 1st attempt at SCAP. Tremory_+_Shuddery.
#Release-1_2: 1st attempt at the 2nd release. Thick_womb_music_cables
#unre[a||eling unstable_conversation_w[g]r[e||app[1]ing.
#Release-1_2_2: The final[ity] + most stab[b{1}ing_with_ur_g(old)athering_eyes]le
#release in the 1.2 series.
#Release-1_3: 1st 1.3 release with [g]host_groin_spas[onic]m[ush]s.
#Release-1_3_2: Latest release with some s[pidery]tone_lizard_clubbed2deathness.
#Release-1_3_3: Latest release with some st[p]arched_+_sw[|t]ollen_body_w[1]ords
#drenched chemically. Yearn_f[l]ingers_cup_sh[g1]immer_throats.


--
```

FIGURE 9. *Screenshot of part 1 of Mez's "_cross.ova.ing][4rm.blog.2.log][_",
written in the pseudo–code language of Mezangelle.*

in bpNichol's 1984 *First Screening,* which I discuss in chapter 2; Mez's unexecutable code poems from the 1990s, written in the fictional programming language Mezangelle (see Figure 9); and Nick Montfort's 2009 open-source Python poetry generator *Taroko Gorge,* which has spawned at least twenty different remixes, present themselves to us as already untied and therefore clearly situated against the sort of black-boxing embodied by the iPad.[64]

Moreover, nearly every early work of digital literature created on the influential hypertext authoring environment Storyspace from the late 1980s through the 1990s is arguably also an instance of hacking in this broad sense. Even though the software—which predates the Web and provides a far richer environment for linking and for linking as mapping than is possible with the one-to-one style of linking that is the basis of the Web—was explicitly created for writers and writing, authors inevitably came up against some feature or even bug they sought to subvert or exploit or felt they needed to create in order to make their text operate in the ways they wanted. For example, Deena Larsen's *Samplers: Nine Vicious Little Hypertexts* from 1997 exploits a bug in Storyspace 1.2C that produces a screen requiring the reader to choose between two writing spaces after they hit Enter (see Figure 10).[65] Larsen writes, "This was crucial in *Samplers,* as I wanted readers to be able to hit enter and see a default story line, but I also wanted readers to be forced to choose at key ventures."[66] In the same work Larsen also takes advantage of the fact that the names of links in *Samplers* can double as phrases that when strung together create what she calls a "shadow story of the main text."[67] Storyspace publisher Mark Bernstein describes how this friendly hack functions:

> Links in Larsen's *Samplers* appear in a dialog box—a conventional list of links that Storyspace authors can use to build an ad hoc multi-tailed link. The dialog is designed to be purely functional, showing a list of links by pathname and

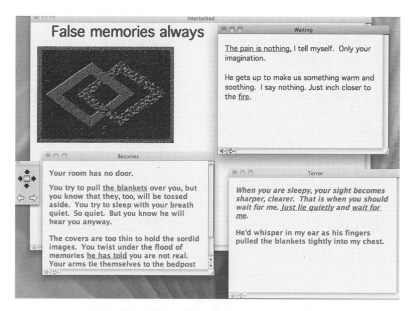

False memories always

The pain is nothing, I tell myself. Only your imagination.

He gets up to make us something warm and soothing. I say nothing. Just inch closer to the fire.

Becomes

Your room has no door.

You try to pull the blankets over you, but you know that they, too, will be tossed aside. You try to sleep with your breath quiet. So quiet. But you know he will hear you anyway.

The covers are too thin to hold the sordid images. You twist under the flood of memories he has told you are not real. Your arms tie themselves to the bedpost

When you are sleepy, your sight becomes sharper, clearer. That is when you should wait for me. Just lie quietly and wait for me.

He'd whisper in my ear as his fingers pulled the blankets tightly into my chest.

FIGURE 10. *Screenshot of Deena Larsen's* Samplers: Nine Vicious Little Hypertexts *from 1997 and the way in which names of links can be strung together to form a secondary but related narrative.*

destination, but Larsen has chosen path names so that this list itself can be read as an interstitial poem.[68]

Finally, Larsen, as well as countless other Storyspace authors, also managed to create the equivalent of easter eggs—called "Jane's Spaces," named after the hypertext literature critic Jane Yellowlees Douglas—in their works. Writes Bernstein in a blog post:

> In hypertext parlance, a Jane's Space is a part of a hypertext that you can't find in the usual, link-following way. A Web page that's not linked to your site and that's hidden from the search engines is a Jane's space; you can only get there if you happen to know the URL. . . . I recently wrote a small program that scans Storyspace documents, looking for spaces with text but no inbound links. Of 28

published hypertexts, at least 16 appear to have Jane's spaces. I knew some of these, but the overall total seems extraordinarily high.[69]

The creation of Jane's Spaces, particularly without the knowledge or the permission of the publisher, is certainly a feat that would not be possible if these authors were creating for Apple multitouch devices, especially given Apple's strict developer guidelines.

Although a handful of digital literature practitioners have found ways to work within and against the strictures of the iPad's tightly controlled hardware and software, since the device is symptomatic of the larger direction in computing toward products that are black-boxed in the name of the supposedly user-friendly, the vast majority of contemporary writers position themselves against the foregoing by using the Web or even Web browsers. By comparison with Storyspace, the Web is certainly more limited, but it is also by far the most profoundly influential and accessible computing platform. Thus, in opposition to the (marketing) rhetoric that celebrates magic, invisibility, seamlessness, and whatever is deemed "natural," work by Talan Memmott, Judd Morrissey, Jason Nelson, and Young-Hae Chang Heavy Industries court glitch on the Web as a way to make the invisible visible once again. Otherwise put, these authors (among numerous others in the field of digital literature) create interfaces that frustrate us as readers, because they seek to defamiliarize the interfaces we no longer notice—a literary strategy akin to Viktor Shklovsky's early twentieth-century invocation of "defamiliarization," which has become the watchword of Russian formalism and its belief about the purpose of art and, by extension, poetic language:

> Art exists that one may recover the sensation of life; it exists to make one feel things, to make the stone stony. . . . The technique of art is to make objects "unfamiliar," to make forms difficult, to increase the difficulty and length of perception

because the process of perception is an aesthetic end in itself and must be prolonged. Art is a way of experiencing the artfulness of an object; the object is not important.[70]

The last line in the foregoing quote is an important point at which digital writers and artists depart, however, from Shklovsky and much of the heritage of the early twentieth-century avant-garde, for these digital writers and artists deploy difficulty and failure to defamiliarize and thus resee interfaces of the present so that we become aware of how the object—in this case, the digital interface—is in fact of utmost importance. Framed as that which gives an account of the normally invisible—the taken-for-granted that nonetheless defines what can be said—the unsettling work by these three authors presents a compelling argument for the importance of digital literature as an intervening force in the computing industry's push to have our devices do all the thinking, perceiving, and even creating for us.

Although *glitch* is rarely used to describe digital literature, the way in which it is commonly used by musicians, gamers, artists, and designers to describe an artistic practice of experimenting with and even aestheticizing the visible results of provoked or unprovoked computer error make it a relevant framework for understanding a whole range of early and contemporary works of difficult digital literature. *Glitch* was first used in the early 1960s to describe either a change in voltage in an electrical circuit or any kind of interference in a television picture. By the 1990s *glitch* broadly described brief bursts of unexpected behavior in electrical circuits, but it was also more specifically used to describe a style of electronic music that was created from already-malfunctioning audio technology (or from causing technology to malfunction) as a way to explore the life of the digital machine and as a reaction against the push in the computing industry to create an ever more clean, noise-free sound. The term has since been appropriated as a name for what Olga Goriumnova and Alexei Shulgin call a "genuine software aesthetics."[71] Glitch aesthetics, then, could include aestheticizing the visible results of a virus or even provoking the computer to take on a virus in order to explore its underlying workings.[72]

Glitch takes this radical shift in what counts as an aesthetic object or an aesthetic experience and asserts that its disruptiveness (in that a glitch constitutes a moment of dysfunctionality in the computer system) defamiliarizes the slick surface of the hardware/software of the computer and so ideally transforms us into critically minded observers of the underlying workings of the computer. As Goriumnova and Shulgin put it, "A glitch is a mess that is a moment, a possibility to glance at software's inner structure. . . . Although a glitch does not reveal the true functionality of the computer, it shows the ghostly conventionality of the forms by which digital spaces are organized."[73] One of the best-known creators of glitch art and games is the Dutch-Belgian collective Jodi, whose members are Joan Heemskerk and Dirk Paesmans. Since the mid-1990s, Jodi has, as they put it in a 1997 interview, battled "with the computer on a graphical level. The computer presents itself as a desktop, with a trash can on the right and pull down menus and all the system icons. We explore the computer from inside, and mirror this on the net. When a viewer looks at our work, we are inside his computer."[74] For example, their 1996 *Untitled Game* is a modification of the video game *Quake* in that the game's architecture no longer functions according to the conventions of gameplay. One way they do this is by exploiting a glitch that is provoked every time the *Quake* software attempts to visualize the cube's black-and-white-checked wallpaper, causing the player to become trapped in a cube.[75] Thus, quite in opposition to the computing industry's attempt to naturalize the interface to the point of invisibility, Jodi makes the interface confusing, unfamiliar, uncomfortable, malfunctioning.[76]

In the field of digital literature, one of the earliest works of glitch is William Gibson's infamous *Agrippa (A Book of the Dead)*, which was published in 1992 as a collaborative effort between

Gibson, book artist Dennis Ashbaugh, and publisher Kevin Begos Jr.[77] It has been thoroughly and subtly discussed by Matthew Kirschenbaum, who understands *Agrippa* in the same terms as all of the works I discuss throughout this book, as exemplifying "the capacity of a digital object to take on and accumulate a material, indexical layer of associations," indicating its own "awareness of the mechanism"—an awareness that ties it to the foregoing hacker-like works of digital literature and that actually reveals itself through its own provoked error.[78] That is, *Agrippa* is packaged as a black box that once opened reveals both a hologram of a circuit board on the underside of the lid and, inside the box, a book, inside of which is nested a 3.5-inch floppy disk that is programmed to encrypt itself after it is used just once. Not surprisingly, once exposed to light, the words and images on the pages of the book fade altogether.[79] Given the self-reflexivity of *Agrippa* and the way its different material components comment on each other, appropriately enough the text of the book doubles as a description of itself and of a photo album that contains fading photographs from the early family history of the narrator, W. F. Gibson Jr.

I hesitated
before untying the bow
that bound this book together.

A black book:
ALBUMS CA. AGRIPPA
Order Extra Leaves By Letter and Name

A Kodak album of time-burned
black construction paper

The string he tied
Has been unravelled by years
and the dry weather of trunks
Like a lady's shoestring from the First World War

Its metal ferrules eaten by oxygen
Until they resemble cigarette-ash
Inside the cover he inscribed something in soft graphite
Now lost
Then his name
W. F. Gibson Jr.
and something, comma,
1924[80]

Agrippa is a work of conceptual writing that not only performs its textual content and itself as a black-boxed black box but also hacks its own mechanism to catalyze its obsolescence and become a book of the dead on, no surprise, dead media.

A more recent instance of digital literature glitch is Talan Memmott's "Lexia to Perplexia" from 2001 (the same year Apple released the iPod, the device with a "yearning toward invisibility" that clearly made way for the iPad)—a work requiring Netscape 4.x or Internet Explorer 4.x to view it such that Memmott quite knowingly built in the work's own protracted, provoked glitch. As every year brings with it the obsolescence of some Web browsers and the updating of others, we slowly lose the ability to access certain parts of "Lexia to Perplexia," if we do not lose the ability to access it altogether. As Memmott writes in the introduction, this work "began as an observation of the fluctuating and ever-evolving protocols and prefixes of internet technology as applied to literary hypermedia. As well, 'Lexia to Perplexia' was originally meant as a critique of both the Author and User/Reader positions in relation to web-based literary content."[81] That is, the reader will notice that in all four sections of the work—"The Process of Attachment," "Double-Funnels," "Metastrophe," and "Exe.termination"—"Lexia to Perplexia" makes wide use of neologisms as a means of presenting, in Katherine Hayles's words, "a set of interrelated speculations about the future (and past) of human-intelligent machine interactions, along with extensive re-inscriptions of human subjectivity and the human body."[82] The text is, however, performed not only linguistically but also narratively and visually. Narratively, Memmott alludes to classical literary references ranging from ancient Greek and Egyptian myth to postmodern literary theory reflecting on humans, technologies, and their collaborative agency. Visually, the work makes use of interactive features that override the source text, leading to a fragmentary reading experience. The functioning and malfunctioning of the interface itself carries as much meaning as the words and the images that compose the text. Memmott instructs his readers to note that the "User/Reader of this piece . . . encounters a number of screens that appear simple upon access. As the User/Reader interacts with the presented objects—images, textual fragments, various UI permutations—the screens are made more."[83] That said, as the years go on, "Lexia to Perplexia" becomes less and less about its linguistic, narrative, and visual elements and more fundamentally about its interface and its slow but sure transformation into an utterly malfunctioning, inaccessible work.

Also published in 2001, Judd Morrissey's "The Jew's Daughter" similarly works against the troubling move toward transparent or invisible computing. In it readers are invited to click on hyperlinks embedded in the narrative text, links that are actually unclickable and that do not lead anywhere so much as they unpredictably change some portion of the text before their eyes.[84] I discuss this work in greater detail in chapter 4 as a way to account for the work's overall complex relationship to the bookbound page—the way in which it reads and reworks both the bookbound page through the digital and the digital through the bookbound page. In the context of this chapter, "The Jew's Daughter" reveals itself as a work that unties the workings of the hyperlinked Web interface, of whose structure we are less and less aware (as we unthinkingly click on any available link on a page) and that more and more seems to be driven by the belief that clicking is an empowering act of identity formation, one that emboldens us to access more-meaningful information and so become active learners and producers of knowledge. In fact, clicking most often simply takes us to something other and yet other again—with most of these clicks carefully monitored by your favorite search engine, which then conveniently sells you back to yourself. Not only has the link become a naturalized structure of the Web, but its very invisibility conceals how our clicks are actually used, nearly always without our awareness.

Likewise working against the clean, "natural," and transparent interface of the Web, Jason Nelson in many of his game poems hybridizes interactive art, video games, and poetry to self-consciously embrace a hand-drawn, handwritten, messy, dissonant aesthetic. In pieces such as the wildly successful "Game, Game, Game And Again Game" from 2007, he also deliberately undoes video game conventions (of accumulation, progress, winning/losing, clear moral victories, immersion) through a nonsensical point system and mechanisms that ensure the most a player ever wins is, for example, a strange home video featuring Nelson playing with action figures in his kitchen (see Figure 11).[85] Nelson has gone on to experiment explicitly with interfaces for digital poetry—creating, in addition to games, everything from mosaic interfaces to cubes, videographs, slot machines, deep-menu poetry, 3D emulations, and circular interfaces. As he states quite unequivocally in an interview with the *Cordite Review*, "Within many digital poems there is one commonality, the emphasis on interface. . . . These interfaces are not just vessels for content, they are poems in themselves. . . . An interface is the life, the body, and a poetic construction in itself."[86]

Finally, although Young-Hae Chang Heavy Industries (YHCHI) are not obviously concerned with either glitch or a hacker aesthetic, insofar as all of their work is defined by a refusal to incorporate interactivity into their works, pieces such as *Traveling to Utopia: With a Brief History of the Technology* use this utter lack of interactivity to create what one might call "clean glitch."[87] This clean-glitch aesthetic is against its own cleanliness in that it uses Adobe Flash to create a spare, mostly black-and-white, cinematic, and totally uninteractive environment

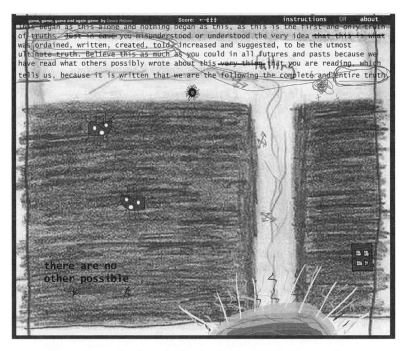

FIGURE 11. *Screenshot from the first level of Jason Nelson's 2007 digital game-poem "Game, Game, Game and Again Game."*

that thereby provides the reader with the ultimate control: to click *away*. They state in an interview from 2005:

> The spectator is far from powerless. She is still the one who decides whether or not she will watch the piece, or having clicked on it, whether she'll click away from it. That's the same power that she has when she considers any other art and literature. Clicking away is one of the essences of the Internet. It's no different from deleting. It's rejection, it's saying "no." That's ultimate power.[88]

Taking a lack of interactivity to such an extreme that it demands spectators reject the work altogether is a gesture that throws them back on themselves and away from the mindless/endless clicking that determines most interactions on the Web.

Consisting of the American poet Marc Voge and Korean artist Young-Hae Chang, who since the beginning of their collaboration in 1999 have written and produced their work in English, French, and Korean, YHCHI is the very definition of unlocatable. Not only does their work slip back and forth between languages, as well as from either a male or a female point of view, but all of their work—whether one calls it net art or digital literature—studiously eschews literary, artistic, and Web conventions. That is, YHCHI intentionally troubles Ezra Pound's dictum to "make it new," which hangs over much twentieth- and even twenty-first-century poetry, creating new works that are new only to the extent that the text and the music is different. Otherwise, every piece that they've created looks identical to every other piece. All of their work begins by mimicking the ten-second countdown that was used by projectionists to focus the film about to be screened. In the same way that the countdown drew the audience's attention to the film as a medium, rather than effacing it altogether as a means to better foster the illusion of film as reality, all YHCHI pieces open with a ten-second countdown that not only alternates between flashing the numeral on the even number (e.g., 10) and the word on the odd (e.g., *nine*) but also ends at *three*, leaving readers/viewers to count down to zero themselves. Similarly, all pieces by YHCHI are marked by the use of a zero instead of the letter O—yet another means by which to force the reader to look at rather than through the text and its interface.

In terms of their disavowal of Web design conventions, their work is created with Adobe Flash simply as a means to present moving, large, bare, black text in Monaco font against a white background (a strategic move against the computing industry's seductive rhetoric perpetually touting the virtues of the new). Pieces by YHCHI are also generally devoid of graphics, colors, photos, illustrations, and interactivity. They write, "We dislike graphic design, and we also dislike interactivity, which are the two staples of web design, if not the web itself. Being artists, we like to do things wrong, or at least our own damn way. We ended up with a moving text synchronized to jazz, which was (and still is) all we could do."[89] Not entirely unlike in Judd Morrissey's "The Jew's Daughter," YHCHI's dislike of interactivity is partly derived from the emptiness of the interactive features in most pieces, which may be touted as offering the reader a liberatory freedom but that in fact simply allow the reader to choose between several predetermined directions. Rather than foster the illusion that their work is an exemplar of democratic literature, they choose to accentuate the absence of freedom in their work. The reader/viewer cannot fast-forward or rewind; they can only click away from the piece and end the experience altogether. YHCHI's dislike of interactivity is also derived from their sense that the Web has become so familiar to us that we're not even aware of its structures, its codes, and the way it works on us rather than us working on it. Distinctly echoing the sentiments of Jodi, they write:

> The Internet and Web have become familiar and even boring and sometimes disagreeable spaces. The Web artist's goal is to make it become less familiar, less boring, less disagreeable, to make it become fresh and new again . . . The computer screen is a superficial support, akin to the surface of a painting. Any Web art that employs images tries to create visual depth to this surface. Any Web art that employs textual information also tries to create depth, albeit with a strategy similar to the writing using: to make the reader forget he or she is looking at ink on a bound page. In this sense, yes, our work and other textual work tries to smash the surface of the computer screen.[90]

While *Traveling to Utopia: With a Brief History of the Technology* has received no critical attention, especially noteworthy in comparison with the broad acclaim given to *Dakota*, it is exemplary of YHCHI's desire to "smash the surface of the computer screen." First, the piece is available in either English/Korean or French/English, and each version is structured slightly differently from the other. The former has large English text in black letters against a white background, with Korean text in green against a black background running across the top of the screen like a stock-market ticker tape and static English text (separate from the main English text at the center of the screen) at the bottom of the screen, also in green against a black background but with a blinking green cursor that's reminiscent of the era of the command-line interface. Already, with only a nod to the visual codes of three different writing interfaces, we have before us a "BRIEF HISTORY OF THE TECHNOLOGY." The French/English version also contains a moving line of green text against a black background on the bottom of the screen, but this time it flashes to the beat of the jazz music playing in the background. The text tells a personal history of the writing technologies that dominated the narrator's life from the time she was a small girl to the time she was an adult, a personal history that is inevitably enmeshed in larger political and national histories. The story begins with the narrator relating her first encounter with a computer, which "LOOKS LIKE A SMALL REFRIGERATOR." She continues, "JUST A GIRL, I THINK ITS MONITOR WITH ITS DIM GLOW, FOR A TV—WEIRD TV SHOW," a naive yet perfectly accurate description of the computer, which has long tried to emulate the TV's ability to masquerade as a window onto an alternate reality.

Immediately after this observation, the narrator provides a description of the only two distinct age markers in her life. The first is the day her father "LEAVES FOR THE MINISTRY AND NEVER COMES HOME. THAT HAPPENS WHEN I'M THIRTEEN." Following this unsettling statement, which leaves the reader wondering whether the father's use of the computer (which he also forbade his daughter to touch, whether for reasons related to her gender or not is unclear) somehow ran counter to the political regime of his day and whether it was connected with his disappearance, the narrator declares:

"WHEN I'M TWENTY I GO ABROAD TO STUDY. I TYPE ON A KEYBOARD WITH A REPEATING SPACE BAR. PRETTY ADVANCED FOR THE PRICE, THE SALESMAN TELLS ME, AND NOT BAD CONSIDERING MOST STUDENTS STILL WRITE THEIR PAPERS BY HAND." Only one or two minutes into this seemingly simple coming-of-age story, we already see how the history of writing technologies is intertwined with surveillance, gender, capitalism, and cultural difference.

As the story unfolds, the narrator then recounts the day that a man appeared at her door, introduced himself as a countryman, and handed her a laptop computer as a gift from "MY LITTLE COMMUNITY." She continues: "BEFORE I CAN RESPOND HE TURNS ON THE COMPUTER'S LITTLE BLACK AND WHITE SCREEN AND SHOWS ME HOW TO USE IT." Then, our visual experience of the piece shifts as the main black text on a white background literalizes the content of the story. At this point the narrator tells of being introduced to fax and e-mail. Recalling the disappearance of her father early on in the story, underlying each introduction of a new writing technology is an ever-present surveillance. First, distant relatives whom she did not inform she had a computer began to call and scold her for not sending them faxes, and then, once she had the ability to e-mail, she was informed by the same countryman that a small fee would be deposited in her account for every e-mail she sent. The narrator ends her story with her return home and her discovery that she now sets off airport security alarms. Shortly after noticing a pain in her abdomen, the narrator came to discover that a Samsung Z-3000 computer chip had been implanted inside her—a computer chip, the text tells us, that is commonly used in global positioning systems or in special collars attached to endangered species for tracking. (Samsung is also, of course, one of the largest Korean-based companies and also claims to have "pioneered the digital age." In the early 1990s they were the largest producer of memory chips in the world.) The piece ends with the narrator claiming she avoids going places where she might set off alarms, staying instead at airport hotels, which are both familiar and exotic—as if she's gone to "A FAR OFF PLACE THAT'S BOTH NOWHERE AND SOMEWHERE."

The meaning of the storyline in *Traveling to Utopia: With a Brief History of the Technology* is as unlocatable as the piece's interface, or its representations of interfaces to comment on interfaces. It is a piece simultaneously of and not of cinema, the Internet, the typewriter, the command-line interface, the windows interface. It is also part fictionalized biography and part allegory for the ways in which access to the contemporary digital world—especially the Web, as I discuss in the postscript—is carefully surveilled and determined by corporations and political maneuverings.

More, alongside work by Memmott, Morrissey, and Nelson, we can read YHCHI's work as a pointed response to the increasing prevalence of invisible interfaces that prevent any kind of making or doing beyond those surface-level activities that are strictly delimited by the interface. With an aesthetic that is either clean or messy, these authors' use of difficulty and defamiliarization by way of digital writing interfaces works against the way in which digital media and their interfaces are becoming increasingly invisible even while these interfaces increasingly define what and how we read/write.

2

From the Philosophy of the Open to the Ideology of the User-Friendly

> In the Old Testament there was the first apple, the
> forbidden fruit of the Tree of Knowledge, which with
> one taste sent Adam, Eve, and all mankind into the great
> current of History. The second apple was Isaac Newton's,
> the symbol of our entry into the age of modern science. The
> Apple Computer's symbol was not chosen purely at random:
> it represents the third apple, the one that widens the paths
> of knowledge leading toward the future.
>
> —Jean-Louis Gassée, *The Third Apple*

Digging to Denaturalize

The second cut into the ground of our technological past in this study of reading/writing interfaces is into the era of the GUI-based personal computer that was preceded by Douglas Engelbart, Alan Kay, and Seymour Papert's experiments with computing and interface design from the mid-1960s to the mid-1970s. This era began with expandable homebrew kits and irrevocably transformed into so-called user-friendly, closed workstations with the release of the Apple Macintosh in late January 1984.[1] Whereas chapter 1 delves into the computing industry's present push to take us more deeply into the era of the interface-free, this chapter uncovers an earlier rupture in the history of computing that partly laid the groundwork for the interface-free.

I look more specifically into the idea that the interface is equal parts user and machine, so that the extent to which the interface is designed to mask its underlying machine-based processes for the sake of the user is the extent to which these same users are disempowered, as they are unable to understand—let alone actively create—using the computer. This chapter concerns itself with a decade in which we can track the shift from a user-friendly computer as a tool that through a graphical user interface (GUI) encouraged understanding, tinkering, and creativity to a user-friendly computer that used a GUI to create an efficient workstation for productivity and task management, as well as the effect of this shift, particularly on digital literary production. Further, the turn from computer systems based on the command-line interface to those based on "direct-manipulation" interfaces that were iconic or graphical was driven by rhetoric that insisted the GUI, particularly that pioneered by the Apple Macintosh design team, was not just different from the command-line interface but *naturally* better, easier, friendlier. As I outline, the Macintosh was, as Jean-Louis Gassée (who headed up its development after Steve Jobs's departure in 1985) writes without any hint of irony, "the *third apple*," after the first apple in the Old Testament and the second apple that was Isaac Newton's, "the one that widens the paths of knowledge leading toward the future."[2] It's worth noting that despite Gassée's hyperbole, which I use to demonstrate the ideological fervor of those working for Apple in the 1980s, his vision for Macintosh was quite different from Jobs's in that Gassée helped shepherd into the market three models of the Macintosh—the Mac Plus, the Mac II, and the Mac SE—that were all expandable, unlike the first-generation Macintosh, which prevented users from opening up the computer by giving them a small electrical shock if they did not adhere to the warnings. (I should point out, however, that the device was not deliberately booby-trapped so much as the Macintosh's power supply required very careful handling, a fact that made it all the more convenient to warn users away from opening it up at all.) While these later models of the Macintosh included expansion slots, which returned Apple philosophically to the era of Steve Wozniak's Apple II—whose eight expansion slots permitted a whole range of display controllers, memory boards, hard disks, etc.—it seems clear that the return of Jobs to Apple in 1997 meant, and continues to mean, a return to keeping the inner workings of Apple computers and computing devices firmly closed off to users. Despite studies released since 1985 that clearly demonstrate GUIs are not necessarily better than command-line interfaces in terms of how easy they are to learn and to use, Apple— particularly, under Jobs's leadership—created such a convincing aura of inevitable superiority around the Macintosh GUI that to this day the same "user-friendly" philosophy, paired with the no longer noticed closed architecture, fuels consumers' religious zeal for Apple products.[3] I have been an avid consumer of Apple products since I owned my first Macintosh PowerBook in 1995. As I write in chapter 1, however, what concerns me is that the user-friendly now takes the shape of keeping users steadfastly unaware and uninformed about how their computers, their reading/writing interfaces, work, let alone how they shape and determine their access to knowledge and their ability to produce knowledge. As Wendy Chun points out, the user-friendly system is one in which users are given the ability to "map, to zoom in and out, to manipulate, and to act," but the result is only a "*seemingly* sovereign individual" who is mostly a devoted consumer of ready-made software and ready-made information to which whose framing and underlying (filtering) mechanisms she or he is not privy.[4]

Thus, the content of this argument is about reversals, and its methodology is defined by tracing the messy, nonlinear rupture I describe in chapter 1—that the shift to the ideology of the user-friendly via the GUI is expressed in contemporary multitouch, gestural, and ubiquitous computing devices, such as the iPad and the iPhone, whose interfaces are touted as utterly invisible and whose inner workings are therefore de facto inaccessible. In this earlier chapter I also outline how this full realization of frictionless, interface-free computing, at least partly born out of the mid-1980s, is in turn critiqued by works of activist digital media poetics.[5]

Using a media archaeology–inspired methodology to understand the historical moment at hand, we can see that activist media poetics played out quite differently in the 1980s than it did in the 1960s' era of the typewriter, as the 1980s was an era newly oriented toward the efficient completion of tasks over and beyond a creative use or misuse of the computer. Arguably, one reason for the heightened engagement in hacking type(writing) in the mid-1960s to the mid-1970s was that the typewriter had become so ubiquitous in homes and offices that it had also become invisible to its users. The point at which a technology saturates a culture is the point at which writers and artists, whose craft is utterly informed by a sensitivity to their tools, begin to break apart that same technology to once again draw attention to the way in which it offers certain limits and possibilities to thought and expression. There are examples of digital poems that inherit this emphasis on making, doing, and hacking, but once again, the vast majority of these works did not appear until both the personal computer and the user-friendly computer whose GUI was designed to keep the user passively consuming technology rather than actively producing it became practically ubiquitous. As I discuss in the following section, activist media poetics in the early to mid-1980s mostly took the form of experimentation with digital tools that at the time were new to writers—an experimentation that at least under the terms set by McKenzie Wark's *Hacker Manifesto,* certainly *could* be framed as hacking. Wark writes that "hackers create the possibility of new things entering the world" and that "the slogan of the hacker class is not the workers of the world united, but the workings of the world untied."[6] As I discuss later in the chapter, works by bpNichol, Geof Huth, and Paul Zelevansky did not make the command-line interface visible so much as they openly played with and tentatively tested the parameters of the personal computer as a still-new writing technology. This kind of open experimentation almost entirely disappeared for several years once Apple Macintosh's design innovations, as well as their marketing, made open computer architecture and the command-line interface obsolete and GUIs pervasive.

Open, Extensible, Flexible: NLS, Logo, Smalltalk

This chapter tackles the notion of the digital interface as a meshing of, even a friction between, human and machine. The degree to which a GUI masks the digital machine for the sake of a more human-like experience is the degree to which users no longer have access to (understanding) both the mechanisms and the flow of information underlying the machine. Likewise, what early human–computer interaction (HCI) designers and researchers struggled with was that the degree to which the interface *unmasks* the digital machine and provides more direct access to the underlying mechanisms is the degree to which it may become more difficult for nonexperts to learn how to use computers.

The interface is only superficially, as Steven Johnson concisely puts it in his canonical *Interface Culture* (1997), "software that shapes the interaction between user and computer . . . a kind of translator" that makes possible the representation of the computer to the user—in the GUI system, through metaphors.[7] As in any translation, there is never a perfect equivalent of the one in the language of the other, and so here, I make clear that we need to think about the nature of computer-to-human translation that takes place via the interface. We can think of the interface as a complex philosophical entity whose translation mechanism is not so much related to natural-language translation as it is to a threshold, along the lines of Matthew Fuller's definition of an interface as containing elements of "the underlying structure of [both] the program and the user."[8] In this way, we can look back and see the philosophy of computing embodied by the early experiments and writing of Douglas Engelbart, Seymour Papert, Alan Kay, and (even) Steve Wozniak as weighted toward a precise midpoint between computer/program and user, a balance that then irrevocably shifted to the user by 1984 with the release of the Apple Macintosh and its icon-based GUI. By contrast, Engelbart, Papert, Kay, and Wozniak show us that a user-friendly computer and graphical interface need not close access to the computer/program for the sake of the user. It can be designed instead to empower users to access and then understand the hardware and the software basics of computing and, ultimately, to create their own tools and applications.

It is, then, not necessarily that the GUI per se is responsible for the creation of Chun's *"seemingly* sovereign individual" but rather that a particular philosophy of computing and design underlying a model of the GUI has become the standard for nearly all interface design. The earliest example of a GUI-like interface whose philosophy was fundamentally different from that of the Macintosh was Douglas Engelbart's oN-Line System (NLS), which he began work on in 1962 and famously demonstrated in 1968 at the Fall Joint Computer Conference in San Francisco. While his "interactive, multi-console computer-display system" with keyboard, screen, mouse, and something he called a chord handset (which allowed the user to issue commands to the computer by pressing different combinations of the five keys) is commonly cited as the originator of the GUI, Engelbart wasn't interested in creating a user-friendly machine so much as he was invested in "augmenting human intellect."[9] As he first put it in 1962, this augmentation meant "increasing the capability of a man to approach a complex problem situation, to gain comprehension to suit his particular needs, and to derive solutions to problems."[10] The NLS was not about providing users with ready-made software and tools from which they chose or consumed but rather about *bootstrapping,* or "the creation of tools for expert computer users," and providing the means for users to create better tools, or tools better suited to their individual needs.[11] In his document editing program, this emphasis on tool building and customization came out of an augmented intellect in Engelbart's provision of view control, which allowed users to determine how much text they saw on the screen, as well as the form of that view—for example, line truncation and content filtering—and of chains of views, which allowed the user to link related files.[12] The NLS's use of view control and chains of view provided a far more direct method of manipulating information—despite the added graphical layer between computer and user—than did the dominant method at that time of punch cards, which very often made the user not a user at all, certainly not one who interacted with the computer but handed over a numerical problem for the computer to solve.[13]

Underlining the fact that the history of computing is resolutely structured by stops, starts, and ruptures rather than by a series of linear firsts, in the year before Engelbart gave his "mother of all demos" Seymour Papert and Wally Feurzeig began work on a learning-oriented programming language called Logo that was explicitly for children but implicitly for learners of all ages. Throughout the 1970s Papert and his team at MIT conducted research with children in nearby schools as they tried to create a version of Logo that was defined by "modularity, extensibility, interactivity, and flexibility."[14] As I discuss briefly in the next section in relation to literary experiments on the Apple II, it was the most popular home computer throughout the late 1970s and until the mid-1980s, and given its open architecture, in 1977 Logo licensed a public version for Apple II computers, as well as for the less popular Texas Instruments TI 99/4. In 1980 Papert published the influential *Mindstorms: Children, Computers, and Powerful Ideas,* in which he makes claims about the power of computers that are startling for a contemporary readership steeped in an utterly different notion of what accessible or user-friendly computing might mean. Describing his vision of "computer-aided instruction" in which "the child programs the computer" rather than one in which the child adapts to the computer or, even, is taught by the computer, Papert asserts that children thereby "embark on an exploration about how they themselves think. . . . Thinking about thinking turns the child into an epistemologist, an experience not even shared by most adults."[15] Two years later, in a February 1982 issue of *Byte* magazine, Logo was advertised as a general-purpose tool for thinking, with a degree of intellectuality rare for any advertisement: "Logo has often been described as a language for children. It is so, but in the same sense that English is a language for children, a sense that does not preclude its being ALSO a language for poets, scientists, and philosophers."[16] Moreover, for Papert, thinking about thinking by way of programming happens largely when the user encounters bugs in the system and has to then identify where the bug is to remove it: "One does not expect anything to work at the first try. One does not judge by standards like 'right—you get a good grade' and 'wrong—you get a bad grade.' Rather one asks the question: 'How can I fix it?' and to fix it one has first to understand what happened in its own terms."[17] Learning through doing, tinkering, experimentation, and trial and error is, then, how one comes to have a genuine computer literacy.

The year after Papert and Feurzeig began work on Logo and the same year as Engelbart's NLS demo, Alan Kay commenced work on the never-realized Dynabook, which was produced as an "interim Dynabook" in 1972 in the form of the GUI-based Xerox Alto, which ran the Smalltalk language. Kay thereby introduced the notion of "*personal* dynamic media" for "children of all ages" that "could have the power to handle virtually all of its owner's information-related needs."[18] Kay, then, along with Engelbart and Papert—all working at the same time, independently yet often influencing each other—very clearly understood the need for computing to move from the specialized environment of the research lab into people's homes by way of a philosophy of the user-friendly oriented toward the flexible production (rather than the rigid consumption) of knowledge.[19] It was a realization eventually shared by the broader computing community, for by 1976 *Byte* magazine was publishing editorials such as "Homebrewery vs the Software Priesthood," which declared, "The movement towards personalized and individualized computing is an important threat to the aura of mystery that has surrounded the computer for its entire history" (see Figure 12).[20] Moreover:

The movement of computers into people's homes makes it important for us personal systems users to focus our efforts toward having computers do what we want them to do rather than what someone else has blessed for us. . . . When computers move into peoples' homes, it would be most unfortunate if they were merely black boxes whose internal workings remained the exclusive province of the priests. . . . Now it is not necessary that everybody be a programmer, but the potential should be there.[21]

It was the potential for programming or, simply, for novice and expert use via an open, extensible, and flexible architecture that Engelbart, Papert, and Kay sought to build into their models of the personal computer to ensure that home computers did *not* become "merely black boxes whose internal workings remained the exclusive province of the priests." As Kay later exhorted his readers in 1977, "Imagine having your own *self-contained knowledge manipulator* in a portable package the size and shape of an ordinary notebook."[22] Designed to have a keyboard, an NLS-inspired chord keyboard, a mouse, a display, and windows, the Dynabook would have allowed users to realize Engelbart's dream of a computing device that gave them the ability to create their own ways to view and manipulate information. Rather than the overdetermined post-Macintosh GUI computer, which has been designed to preempt each user's every possible need through the creation of an overabundance of ready-made tools and whose underlying workings are now utterly black-boxed such that, as homebrewers protested in the mid-1970s, "those who wish to do something different will have to put in considerable effort," Kay wanted a machine that was "designed in a way that any owner could mold and channel its power to his own needs . . . a metamedium, whose content would be a wide range of already-existing and not-yet-invented media."[23] More, Kay understood from reading Marshall McLuhan's *Understanding Media* that the design of

Personal computing people stand to be largely independent of the priesthood because they are strikingly sophisticated and because they freely share their ideas. A very good example of both these traits can be found in the nearly spontaneous generation of Tiny BASIC through the medium of the People's Computer Company and Dr Dobb's *Journal of Computer Calisthenics and Orthodontia*. One issue published some rough design notes for a machine independent Tiny BASIC, but that was only the beginning. The next few issues published refinements on the design and later ones included an implementation in an interpretive language and then both octal and annotated source programs realizing the interpreter and the entire system in 3 K of 8080 code. To top it off, the whole project was done by far-flung individuals in less than a year.

While Tiny BASIC is a very striking example of what amateurs can do when they work together, we cannot afford to ignore its extreme dependence on good fortune to bring it to pass. Your own copy of BYTE magazine is another example; it is the result of one man's frustration at making his own computer work and his desire to let others profit by his experience. We've been very lucky to have a few people with high ideals to point the way for us, but we would be ill advised to depend on having these fortunate circumstances continue. The time is ripe for the community of personal computing enthusiasts to start thinking seriously about supplying its own steam to back up the energies put out by a few people with strong motivations to help launch the personal computing movement. It's launched now, and we have to provide the impetus and direction to make sure it develops in a way beneficial to the community at large.

A good example of a means to distribute software which divides the effort fairly and in a way nobody seems to mind is the software exchange of the Homebrew Computer Club in the San Francisco Bay Area. At each meeting (every two weeks) there is a table covered with paper tapes of programs contributed by all and sundry. Anybody is welcome to take any tape at all, subject only to the proviso that each copy taken from one meeting be replaced by at least one

91

FIGURE 12. *An image from the editorial "Homebrewery vs the Software Priesthood" that appeared in* Byte *magazine in October 1976.*

this new metamedium was no small matter, for the use of a medium changes an individual's and a culture's thought patterns.[24] Clearly, Kay wanted thought patterns to move toward a literacy that involved reading and writing in the new medium instead of the unthinking consumption of ready-made tools, for he wrote, "The ability to 'read' a medium means you can *access* materials and tools created by others. The ability to 'write' in a medium means you can *generate* materials and tools for others. You must have both to be literate."[25]

While Kay envisioned that the GUI-like interface of the Dynabook would play a crucial role in realizing this metamedium, the Smalltalk software driving this interface was equally necessary. Its goal was, as the principal designer, architect, and implementer Daniel Ingalls wrote in a 1981 special issue of *Byte* dedicated to Smalltalk, "to provide computer support for the creative spirit in everyone."[26] While 1971 was the year Alan Kay's Learning Research Group at Xerox PARC developed a working version of Smalltalk—also introducing for the first time, via Smalltalk-71, the term *object oriented programming* (OOP), a paradigm now supported by nearly all modern programming languages—1980 was the year Kay's group released Smalltalk-80 to the public, a version that was then featured in the aforementioned issue of *Byte*. Although examining how the workings of Smalltalk and OOP manifested their overarching philosophy is important, my interest is in tracking this philosophy as part of a broader trend in computing from the 1970s until the mid-1980s—one that is reflected largely in the discourse around GUIs and the user-friendly. Those who worked on Smalltalk saw it as a fundamental break from the philosophy of the closed, elitist, decidedly undemocratic "software priesthood." Not surprisingly, Kay and his collaborators began working intensely with children after the creation of Smalltalk-71. Influenced by developmental psychologist Jean Piaget, as well as Kay's own observation of Papert and his colleagues' use of Logo in 1968, Smalltalk relied heavily on graphics and animation through one particular incarnation of the

GUI: the Windows, Icons, Menus, and Pointers (WIMP) interface. Kay writes that in the course of observing Papert using
Logo in schools, he realized that these children were "doing real
programming":

> This encounter finally hit me with what the destiny of
> personal computing *really* was going to be. Not a personal
> dynamic *vehicle,* as in Engelbart's metaphor opposed to the
> IBM "railroads", but something much more profound: a per-
> sonal dynamic *medium.* With a vehicle one could wait until
> high school and give "drivers ed", but if it was a medium, it
> had to extend into the world of childhood.[27]

As long as the emphasis in computing was on learning—
especially through making and doing—the target demographic
was going to be children, and as long as children could use the
system, then so too could any adult, provided they understood
the underlying structure, the how and the why, of the programming language. As Kay astutely remarks, *"We make not just to
have, but to know.* But the having can happen without most of
the knowing taking place."[28] As he goes on to point out, designing the Smalltalk user interface shifted the purpose of interface design from "access to functionality" to an "environment
in which users learn by doing."[29]

Smalltalk designers did not completely reject the notion of
ready-made software so much as they sought to provide users
with a set of software building blocks that they could combine
and/or edit to create their own customized systems. As Trygve
Reenskaug, a visiting Norwegian computer scientist with
the Smalltalk group at Xerox PARC in the late 1970s, put it:

> The new user of a Smalltalk system is likely to begin by
> using its ready-made application systems for writing and
> illustrating documents, for designing aircraft wings, for
> doing homework, for searching through old court decisions,

for composing music, or whatever. After a while, he may become curious as to how his system works. He should then be able to "open up" the application object on the screen to see its component parts and to find out how they work together."[30]

With an emphasis on learning and building through an open architecture, Adele Goldberg—codeveloper of Smalltalk along with Alan Kay and author of most of the Smalltalk documentation—describes, in the 1981 special issue of *Byte,* the Smalltalk programming environment as one that sets out to defy the conventional software-development environment (see Figure 13).

In Figure 13, the Taj Mahal in the left-hand Figure 1 "represents a complete programming environment, which includes the tools for developing programs as well as the language in which the programs are written. The users must walk whatever bridge the programmer builds."[31] By contrast, the right-hand Figure 2 represents a Taj Mahal in which the "software priest" is transformed into one who merely provides the initial shape of the environment, which programmers can then modify by building "application kits" or "subsets of the system whose parts can be used by a nonprogrammer to build a customized version of the application."[32] The user or nonprogrammer is, then, an active builder in a dialogue with the programmer instead of a passive consumer of a predetermined and, perhaps, overdetermined environment.

At roughly the same time as Kay began work on Smalltalk in the early 1970s, he was involved with the team of designers working on the NLS-inspired Xerox Alto, which was developed in 1973 as, again, an "interim Dynabook" that had a three-button mouse and a GUI working in conjunction with the desktop metaphor and that ran Smalltalk. While only several thousand noncommercially available Altos were manufactured, its GUI and its network capabilities, as team members Chuck Thacker and Butler Lampson believe, made it quite likely the

Figure 1   Figure 2

part of Alan Kay's personal computing vision, the Dynabook. The vision is a hand-held, high-performance computer with a high-resolution display, input and output devices supporting visual and audio communication paths, and network connections to shared information resources. LRG's goal is to support an individual's ability to use the Dynabook creatively. This requires an understanding of the interactions among language, knowledge, and communication. To this end, LRG does research on the design and implementation of programming languages, programming systems, data bases, virtual memories, and user interfaces.

The ivory tower on the island of Smalltalk is an exciting, creative place in which to work on these ideas. A

sense of LRG's long-range goals is aptly portrayed in the illustrations designed by Ted Kaehler.

In figure 1, we see a view of the conventional software development environment: a wizard sitting on his own computational cloud creating his notion of a Taj Mahal in which programmers can indulge in building applications for nonprogramming users. The Taj Mahal represents a complete programming environment, which includes the tools for developing programs as well as the language in which the programs are written. The users must walk whatever bridge the programmer builds.

A goal in the design of the Smalltalk system was to create the Taj Mahal so that programmers can modify it by building *application kits*, which are specialized exten-

FIGURE 13. *Image by Adele Goldberg in a special issue of* Byte *magazine from 1981 on Smalltalk in which she contrasts the conventional philosophy of software driven by "wizards" (Figure 1) and that provided by Smalltalk for the benefit of the programmer/user (Figure 2).*

first computer explicitly called a "personal computer." By 1981 Xerox had designed and produced a commercially available version of the Alto called the 8010 Star Information System, which was sold along with Smalltalk-based software. As Jeff Johnson et al. point out, the most important connection between Smalltalk and the Xerox Star lay in the fact that Smalltalk could clearly illustrate the compelling appeal of a graphical display that the user accessed via mouse, overlapping windows, and icons (see Figure 14).[33]

The significance of the Star for this chapter is, however, partly the indisputable impact it had—or rather, Smalltalk had—on the GUI design of first the Apple Lisa and then the Macintosh. Its significance is also in the way it was labeled clearly as a workstation for "business professionals who handle information" rather than as a metamedium or as a tool for creating or for thinking about thinking that could be encompassed by the term *workstation,* as we can see in Douglas Engelbart's definition of it as a "portal into a person's 'Augmented Knowledge Workshop'—the place in which he finds the data and tools with which he does his knowledge work."[34] But the Star's interface, which was the first commercially available computer born out of work by Engelbart, Papert, and Kay that attempted to satisfy both novice and expert users in providing an open, extensible, flexible environment and that also happened to be graphical, was conflicted at its core. While in some ways the Star was philosophically very much in line with the open thinking of Engelbart, Papert, and Kay, in other ways its philosophy as much as its GUI directly paved the way to the closed architecture and consumption-based design of the Macintosh.

Take, for example, the overall design principles of the Star, which were aimed at making the system seem "familiar and friendly." Designers David Canfield Smith, Charles Irby, Ralph Kimball, and Bill Verplank avowed, in a 1982 special issue of *Byte,* to avoid the characteristics listed on the right while adhering to a schema that exemplified the characteristics listed on the left:

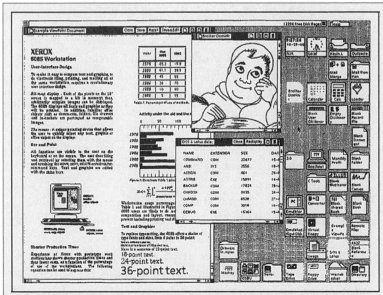

Figure 1. ViewPoint screen image. Star's bitmapped display, once unique in the marketplace, is now much more common. Such a display permits WYSIWYG editing, display of proportionally spaced fonts, integrated text and graphics, and graphical user interfaces.

windows and then cut and pasted together. For example, a MacDraw drawing put into a Microsoft Word or Aldus Pagemaker document can no longer be edited; rather, the original must be re-edited with MacDraw and then substituted for the old drawing in the document.

Not even Star is fully integrated in the sense used here. For example, though the original structured graphics editor, the new one (see "History of Star development" below), and the table and formula editors all operate inside text files, spreadsheets and freehand drawings are currently edited in separate application windows and transferred into documents, where they are no longer fully editable.

User-interface level. Star's user interface is its most outstanding feature. In this section we discuss important aspects of the interface in detail.

Desktop metaphor. Star, unlike all conventional systems and many window- and mouse-based ones, uses an analogy with real offices to make the system easy to learn. This analogy is called "the Desktop metaphor." To quote from an early article about Star:

Every user's initial view of Star is the Desktop, which resembles the top of an office desk, together with surrounding furniture and equipment. It represents a working environment, where current projects and accessible resources reside. On the screen are displayed pictures of familiar office objects, such as documents, folders, file drawers, in-baskets, and out-baskets. These objects are displayed as small pictures, or icons.

The Desktop is the principal Star technique for realizing the physical office metaphor. The icons on it are visible, concrete embodiments of the corresponding physical objects. Star users are encouraged to think of the objects on the Desktop in physical terms. You can move the icons around to arrange your

Desktop as you wish. (Messy Desktops are certainly possible, just as in real life.) You can leave documents on your Desktop indefinitely, just as on a real desk, or you can file them away.[1]

Having windows and a mouse does not make a system an embodiment of the Desktop metaphor. In a Desktop metaphor system, users deal mainly with data files, oblivious to the existence of programs. They do not "invoke a text editor," they "open a document." The system knows the type of each file and notifies the relevant application program when one is opened.

Most systems, including windowed ones, use a Tools metaphor, in which users deal mainly with applications as tools. Users start one or more application programs (such as a word processor or spreadsheet), then specify one or more data files to edit with each. Such systems do not ex-

FIGURE 14. *Screenshot of the Xerox Star desktop that appeared in Jeff Johnson et al.'s "The Xerox Star: A Retrospective" in a 1989 issue of* Computer.

Easy	*Hard*
concrete	abstract
visible	invisible
copying	creating
choosing	filling
recognizing	generating
editing	programming
interactive	batch[35]

While there's little doubt that ease of use was central to Engelbart, Papert, and Kay—often brought about through interactivity and making computer operations and commands visible—the avoidance of "creating," "generating," and "programming" could not be further from their vision of the future of computing. This divided loyalty to two different notions of the user-friendly was more specifically exemplified by the Star's system of commands. Rather than typing out a command from memory via the command-line interface or, even, selecting a command from a menu, commands on the Star took the form of icons that functioned, as the designers describe it, as both noun and verb. The noun was whatever object on the screen the user wished to manipulate, whether file or document or application, and the verb was the type of action or manipulation the user wished to perform. Selection took place by the user hovering the cursor over the object, clicking the mouse button to select the action, and then hitting the Next key on the keyboard to select content from the next field in the document. Other commands that appeared on the keyboard included Find, Open, and Close. Curiously, the designers' explanation of Star's commands ends with the declaration, "Since Star's generic commands embody fundamental underlying concepts, they are widely applicable. . . . Few commands are required. This simplicity is desirable in itself, but it has another subtle advantage: *it makes it easy for users to form a model of the system. What people understand, they can use.*"[36] In other words, at the same time the

Star precluded creating, generating, and programming through its highly restrictive set of commands in the name of simplicity (restrictions that most certainly excluded certain creative possibilities), it also wanted to promote users' understanding of the system as a whole—although again, this particular incarnation of the GUI represented the beginning of a shift toward only a superficial understanding of the system. Without a fully open, flexible, and extensible architecture, the home computer became less a tool for learning and creativity and more a tool for simply "handling information."

Writing as Tinkering: The Apple II and bpNichol, Geof Huth, and Paul Zelevansky

We can clearly see this shift from the philosophy of the open to the ideology of the user-friendly work machine not only in the structure of Steve Wozniak's Apple II versus Steve Jobs's Apple Macintosh but also in the utterly different marketing strategies for these two machines. Wozniak's Apple II used a command-line interface instead of a GUI and was aligned philosophically with homebrewery in that its eight expansion slots allowed users to add on a range of devices, including display controllers, memory boards, and hard disks, which meant its open architecture was explicitly for tinkering and, thus, creativity. Writing for *Byte* in May 1977, the month before the public release of the Apple II, Wozniak declared:

> I designed the Apple-II to come with a set of standard peripherals, in order to fit my concept of a personal computer. In addition to the video display, color graphics and high resolution graphics, this design includes a keyboard interface, audio cassette interface, four analog game paddle inputs . . . three switch inputs, four 1 bit annunciator outputs, and even an audio output to a speaker. Also part of the Apple-II design is an 8 slot motherboard for IO.[37]

In the months leading up to its release, the Apple II was advertised as not only a task-management machine but also a means for imagination and invention:

> You can use your Apple to analyze the stock market, manage your personal finances, control your home environment, and to invent an unlimited number of sound and action video games. That's just the beginning. . . . *You don't want to be limited by the availability of pre-programmed cartridges. You'll want a computer, like Apple, that you can also program yourself.* . . . The more you learn about computers, the more your imagination will demand. So you'll want a computer that can grow with you as your skill and experience with computers grows. Apple's the one.[38]

Eight months later, in November 1977, Apple even issued a contest for "the most original use of an Apple since Adam," with creative use in near diametrical opposition to Gassée's later framing of the Macintosh as a computer that was, in and of itself and regardless of use, the most original "apple" since Newton.

Not surprisingly, then, the Apple II was by the early 1980s the first home computer that appealed to writers looking to experiment with this new medium of expression—writers who were keen to take up John Cage's injunction from 1966 to use a computer not as a labor-saving device but rather as one that increased work for the writer mostly insofar as the computer's graphical, algorithmic, and interactive capabilities encouraged experiments with form.[39] It also made sense that writers chose the Apple II over other available home computers of the time. Even though sales of the Apple II were initially slow in comparison with those of the much less expensive Commodore PET and TRS-80 Model 1, by 1981, once Apple had added their floppy-drive accessory and then both started a highly effective ad campaign (claiming that it was the "best-selling personal computer") and created the first-ever spreadsheet application,

VisiCalc, the Apple II was the best-selling personal computer.[40] Just two years later, in 1983, Apple released arguably its best-selling computer, the IIe, which crucially for writers not only allowed uppercase and lowercase letters but also had an eighty-column display, in contrast with the first-generation Apple II, which was uppercase only and had a forty-column display.

Canadian experimental writer bpNichol not only promptly purchased an Apple IIe the year it was released but also began work on one of the first published works of digital literature, *First Screening,* a series of twelve kinetic poems written in the Apple BASIC programming language (see Figure 15).[41] Given his typewriter-based experiments with highly visual, permutational, DIY-oriented, and processual concrete poems, which I discuss in chapter 3, coupled with his McLuhan-inspired understanding of writing tools as extensions of the writer, it is not surprising that Nichol's writing experiments extended to the computer, exploited the possibilities of a screen-based medium, and so resulted in the creation of these twelve kinetic, cinematic poems. In fact, as Nichol acknowledges in the accompanying printed matter from 1984, he was surprised that in the process of composing *First Screening* in BASIC,

> concerns that had been present for me in the mid-60s, issues of composition and content i was confronting while working with my early concrete poems, suddenly found a new focus. In fact, i was finally in a position to create those filmic effects that i hadn't had the patience or skill to animate at that time. . . . Computers & computer languages also open up new ways of expressing old contents, of revivifying them. One is in a position to make it new.[42]

Because the poems in *First Screening* move soundlessly across a black computer screen, the work is new in how it positions itself halfway between film and sound/concrete poetry and self-consciously (mis)uses the filmic medium to create one of the first kinetic digital poems.

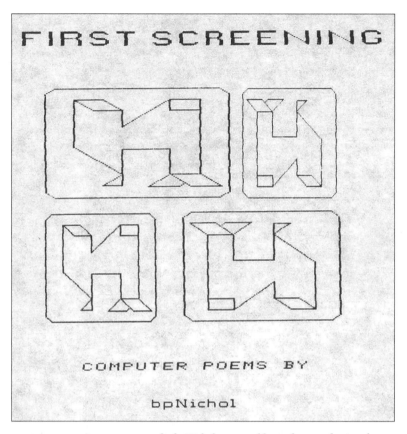

FIGURE 15. *A concrete poem by bpNichol presumably made on and printed from his Apple IIe. The poem appears on an insert that came with the 5.25-inch floppy for his 1983–84* First Screening. *Reprinted by permission of Eleanor Nichol on behalf of the bpNichol Estate.*

In *First Screening* it appears as though Nichol—writing at the very beginning of the era of the personal computer—understands the ease with which the digital computer has an entirely different effect on the body than that of a reading/writing machine such as the typewriter. For example, midway through the screening, the reader/viewer is introduced to "ANY OF YOUR LIP: a silent sound poem for Sean O'Huigin." The title of this piece alone gestures to the absent presence of the body.

Once the poem begins, we see/read the kinetic permutations that move between "MOUTH" and "mouth," "myth" and "MOUTH," "math" and "MOUTH," "mate" and "MOUTH," "maze" and "MOUTH," and "amaze" and "MOUTH" and then the alternation between "ing," "amaze," and "MOUTH," which closes with the repeated flashing of "ing" and, finally, "MOUTH." That said, while the poem is perhaps silent because of the limits of Nichol's own programming know-how (not to mention the limited sound capabilities of the Apple IIe itself), it is noticeable how this paradoxically silent sound poem draws attention to its silence at the same time it enacts and perhaps even encourages readerly interactivity. Especially with the repeated flashing of "ing" at the end of the poem, a verb ending that signals generalized or uncompleted action, "ANY OF YOUR LIP" invites readers to sound out or to "mouth" the words as they try to make sense of the connections between the words while they flash across the screen.

The poems in *First Screening* are not interactive in the sense to which we are accustomed, and the underlying code of the poems shows an iteration of interactivity that does not depend on clicking links. Looking at the BASIC code gives a clear sense of the permutational nature of the kinetic poems, as Nichol carefully moves each letter up and down the vertical axis through the VTAB command and across the screen with HTAB. For example, the following are the first four lines of code for the poem "SAT DOWN TO WRITE YOU THIS LETTER":

```
640 VTAB 12: HTAB 5: PRINT "AT DOWN TO WRITE YOU
    THIS POEM S"
645 VTAB 12: HTAB 5: PRINT "T DOWN TO WRITE YOU THIS
    POEM SA"
650 VTAB 12: HTAB 5: PRINT "DOWN TO WRITE YOU THIS
    POEM SAT"
655 VTAB 12: HTAB 5: PRINT "DOWN TO WRITE YOU THIS
    POEM SAT"
```

The fourth line of code then prints "DOWN TO WRITE YOU THIS POEM SAT" on-screen (see Figure 16).

More, as Jim Andrews astutely discovered in the process of putting *First Screening* online, the twelfth poem does not appear on-screen, as it is instead nested in the last eight lines of the code—a poem that is also one of the first works of codework, or literary writing that is code but not necessarily executable.[43] A reader would discover this piece only if she or he understood the underlying workings of the poem, rather than simply taking in its on-screen effects, and noticed that on line 116 was a REM (or remark, a way of leaving explanatory comments in the code) that states, "FOR FURTHER RE-MARKS LIST 3900,4000." Ideally, this statement would prompt the curious reader to type, "LIST 3900,4000," and view the following further "RE-MARKS":

```
3900 REM ARK
3905 REM BOAT
3910 REM AIN
```

FIGURE 16. *Screenshot of an emulated version of what appears on-screen as a result of line 650 of the Apple BASIC code of bpNichol's* First Screening.

```
3915 REM RAIN RAIN RAIN RAIN RAIN RAIN RAIN RAIN
     RAIN RAIN RAIN RAIN RAIN RAIN RAIN RAIN RAIN
     RAIN RAIN RAIN RAIN RAIN RAIN RAIN RAIN RAIN
     RAIN RAIN RAIN RAIN RAIN RAIN RAIN RAIN RAIN
     RAIN RAIN RAIN RAIN RAIN
3920 REM BOAT
3925 REM ARK
3930 REM BOW
3935 REM ARC
4000 END
```

Nichol begins his permutational concrete poem by breaking apart "REMARK" to form "REM" and "ARK," which is followed by "REM BOAT" to make sure we understand this *ark* is not only biblical—rather than the French-derived bow, sometimes spelled *arc*—but also a reference to Noah's ark, not the Ark of the Covenant. Continuing his permutational punning, "RE-MARK" is then turned into "REM AIN," the remains of which produce forty appearances of the word "RAIN." After leaving "ARK," "BOW" appears as an "ARC" across the sky that is, this time, a symbol of the promise God made with Noah to never again flood the earth. This work is, again, not an example of activist media poetics in the sense that becomes more prevalent once the model of the closed computer with an invisible GUI is ubiquitous but rather, given the homebrew-inspired open architecture of the Apple II, of writing as DIY tinkering.

First Screening was influential enough among experimental writers of the time that a few years later, in 1987, Geof Huth produced "Endemic Battle Collage"—what he called "aural and kinetic poems"—for the Apple IIe, in the tradition of Nichol's earlier kinetic/permutational poems (see Figure 17).[44] Huth clearly identified with Nichol's tinkering with the limits and possibilities of writing media. He writes, "My work is based on thinking about what new tools to use and what new possibilities to achieve. bpNichol, the poet I most identify with, seemed to me a poet who understood this and practiced this himself." A fundamental difference between "Endemic Battle Collage" and *First Screening* is, however, that the former demonstrates a significantly more sophisticated understanding of Apple BASIC in that it incorporates color and sound. The letters twirl, spin, and rotate in more complex patterns, and Huth plays with a system of highlighting words on-screen as a way to play with white text on a black background and black text on a white background. It is also worth noting that by 1987 the Apple Macintosh had been available for three years, yet Huth still chose to experiment with the Apple IIe. Other than the fact, I would argue, that the IIe was a more appropriate machine for literary experimentation, it was also significantly less expensive and so more appealing to writers and artists, for in 1987 one could purchase a IIe for $1,400, whereas a Macintosh retailed for about $2,500.

Given this connection between those who sought to extend their formal and medium-specific writing experiments to the Apple II and its predecessors, it is no coincidence that those working with artists books (a genre concerned with playing with material dimensions and with conventions of the book as a technology) also looked to the Apple II as a means to create a new form that was a hybrid nestled between the computer and the book. In 1986 Paul Zelevansky published the second volume of his by now rare artist book trilogy *The Case for the Burial of Ancestors*. This second volume, *Book Two: Genealogy,* is supposedly the third edition (which is a fiction, since there was only one edition) of a fictional translation of an equally fictional ancient text, which is itself a translation of an oral account of the Hegemonians from the twelfth and thirteenth centuries that is "attributed to a score of mystics, religionists and scholars, none of whom has ever stepped forward."[45] The text focuses particularly on the stories of four priests, each of whom is identified throughout the book with a different typeface, which Zelevansky claims makes it possible "to build a reading of the text around a typographical sequence."[46] Also included in *Book Two* is a sheet of sixteen stamps—each a miniature, layered collage of letters and found objects. As Zelevansky puts it in "Preface to

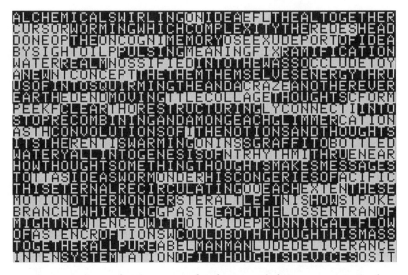

FIGURE 17. *Screenshot from an emulated version of a later sequence in Geof Huth's 1987 Apple BASIC poem "Endemic Battle Collage."*

the Third Edition," "Each stamp has a particular part to play in the narrative. It is left to the Reader to attach them, where indicated, in the spaces provided throughout the text."[47] Finally, enclosed in an envelope on the inside of the back cover, the book comes with "SWALLOWS," a 5.25-inch floppy disk containing a video game that forms the first of the book's three parts. Programmed in Forth-79 for the Apple IIe or II+, the original "SWALLOWS" was supposedly created in 1985 and integrated into the first part of the print version of *Book Two* through a short text and image version.[48] Further, not only are the separate parts of *Book Two* tightly intertwined with each other, but so too are the first and second books of the trilogy, for the imagery and marginalia in the book itself are, we are led to believe, all drawn either from *Book One* or from the "SWALLOWS" disk:

1. Graphic symbols (black on white) based on those found in the second edition of GENEALOGY and in Book I. of THE CASE FOR THE BURIAL OF ANCESTORS.

2. Facsimiles of computer imagery (white on black) drawn from SWALLOWS.

3. Computer printout (black on white) sent ("dumped") from SWALLOWS to a dot-matrix printer.[49]

Not surprisingly, despite Zelevansky's claims that certain facsimiles and computer printouts in the "SWALLOWS" section are taken from the game, the book version includes images and text that do not in fact appear in "SWALLOWS," although it's possible they did exist in the original 1985 version of the game, though any notion of an original in the trilogy should be considered with a degree of skepticism. Appropriately enough, the opening text, "I. HOW IT BEGAN"—the first of eleven parts in the book, which of course correspond only occasionally to the seven parts in the game—is preceded by an image that *does not* appear in the game and begins with an excerpt that *does* appear in the opening scenes of the game:

SITTING ABOVE THE ACTION, PULLING STRINGS,
IT WAS THE PUPPETEER'S GAME TO PLAY—
AND THE PUPPETEER LIKED TO PLAY.

THE KNOWN WORLD WAS IN PLACE.
THE KNOWN PEOPLE WERE IN MOTION.
WHAT ELSE WAS NEW?
THE PUPPETEER DEMANDED SOMETHING MORE.[50]

Thus, in both the book and the game, nearly all of the subjects—whether priest, swallow, or puppeteer—could also be stand-ins or even allegories either for each other or for us as readers/players. At a minimum this passage seems to frame "SWALLOWS" as a game about us playing the game as we sit "ABOVE THE ACTION, PULLING THE STRINGS" by navigating our way through, choosing menu options, and progressing from one level to another. And the "SOMETHING MORE" alluded to at the end of the passage? It is at least partly the unlocatable nature of the text, of

whose meaning, structure, origin, authorship, and even boundaries we can never be certain. As Zelevansky writes in the following lines, which appear only in the book and not in the game:

> AFTER ALL . . .
> WHO WAS THE RULER OF THE TENT?
> WHO WAS THE OWNER OF THE WORK?
> WHO WAS MINDING THE STORE?
> WHO WAS KEEPING THE SCORE?
> SUDDENLY, LIGHTNING STRUCK THE DINNER TABLE,
> THE DIRECTOR BROUGHT FORTH SPECIAL EFFECTS
> AND THE GAME OF SWALLOWS WAS BORN.
> IT WAS ELECTRONIC MYTHOLOGY FOR AN AUDIENCE
> OF ONE—IN HONOR OF THE PUPPETEER.[51]

At this point, most of the text in the "SWALLOWS" section of
the book recaps or sometimes replicates certain parts from the
story, on the disk, of four swallows (which, again, could stand in
for the four priests or could stand in for our experience reading/
interacting with the trilogy as a whole) who lose their way home
because of "SOME SMOKE." Once the smoke clears, the swallows discover they cannot "FIND THE ORIGINAL" home, and
so they go about trying to rebuild another home. Eventually,
the swallows find they can no longer avoid a looming existential question: "WHO ARE THEY ANYWAY?" The answer that
appears in both the book and the game is as follows:

> IT HAD TO BE FACED, THE SWALLOWS WERE
> EXPENDABLE. THE MONITOR ATE THEM WITH
> GREAT REGULARITY.
> WAS THIS FAIR? DID THE PUPPETEER CARE?
> DOES A MACHINE KNOW IT'S ACTING LIKE A
> MACHINE?[52]

The existential question looming over the swallows now looms
over us as players/readers, as well as the machine. If we are the

puppeteer, do we in fact care about what happens on-screen? Is it possible that the machine knows, let alone cares, what happens in the machine and on-screen? And further, Zelevansky urges us "TO CONSIDER THIS":

> THOUSANDS OF ELECTRONIC SIGNALS SETTING
> OFF THOUSANDS OF FLASHING LIGHTS, PRODUCED
> THE VISIBLE EFFECTS OF THIS GAME.
> AS WITH THE SWALLOWS, EACH LIGHT FOLLOWS
> A PATH THROUGH THE GAME; EACH LIGHT HAS
> ITS BEGINNING AND ITS END, ITS HOME AND A
> MULTITUDE OF POSSIBLE DESTINATIONS.[53]

This passage is just one of many examples of how the game self-reflexively talks about itself both as a fictional construct and as nonfiction, insofar as the game is mediated and structured specifically by the computer. Further, once the reader/player interacts with the game "SWALLOWS," some of these "POSSIBLE DESTINATIONS" include nine choices offered via "CAMEL MENUS" that appear in four of the seven parts of the game: "F TO FLY," "B TO BUILD," "W FOR CAMEL WISDOM," "D FOR DIVINE INTERVENTION," "Q FOR OLD QUESTIONS," "ESC TO RETURN TO THE BEGINNING," "& FOR THE NEXT CHAPTER," "P FOR PAST CHAPTER," and "R FOR RANDOM FLIGHT." These choices are not only decidedly unconventional for a video game but also foreshadowed in the game's text such that the narrative seems to be aware of itself rather than the reader/player being the sole owner of an awareness that is usually structurally reinforced in a game because of the separation that exists between the story and the game controls.

Last, if you thought "SWALLOWS" couldn't remix itself any more or recede any more from the present moment as a result of obsolescence, thanks to Matthew Kirschenbaum and the expertise and resources at the Maryland Institute for Technology in the Humanities, in 2012 Zelevansky resurrected "SWALLOWS" by first creating a disk image and then an emulation of the original. From there, Zelevansky was able to free himself from his slowly ailing Apple computer and go on to create "SWALLOWS 2.0"—a movie we can watch and download. "SWALLOWS 2.0" is, in his words, "a conversation between an Apple IIe, and a Macbook Pro" and is yet another self-conscious, self-referential remix of the "original" version that makes it clear we are watching an emulated, thoroughly mediated version that includes additional audio, video, and even fake sequences from the Apple IIe that masquerade as pieces from the so-called original "SWALLOWS." It is as if "SWALLOWS 2.0" acts out the story of the swallows from the text of the book, who, again, find they cannot "FIND THEIR ORIGINAL" home and need to rebuild another, for without using the 5.25-inch floppy itself along with an Apple IIe or II+, there is no original "SWALLOWS." It's remix all the way down.

In short, then, "SWALLOWS," or even *Book Two* as a whole—more so than *First Screening* and "Endemic Battle Collage"—is a very early literary instance of a work that self-consciously uses its own text, distributed across different media, to comment on these media and on the nature of our interactions with the text as it is mediated by these particular reading/writing technologies—whether book, video game, or stamp. It also thereby works against the grain of each medium to accomplish this level of metacommentary, leading the way quite clearly to later works of activist media poetics that seek to make visible once again the underlying workings of the computer and the digital interface lying at the threshold of the user and the computer.

Closed, Transparent, Task Oriented: The Apple Macintosh

Where are the works of digital literature created for the Apple Macintosh, the successor to the Apple II line of computers? I would say that if they do exist (most likely, a number of early Storyspace works were created on the Macintosh), then they do so in spite of the Macintosh, a computer clearly designed for consumers, not creators. As seen in the advertisements in Figures 18 and 19, it was marketed as a democratizing machine when in fact it was democratizing only insofar as it marked a profound shift in personal computing away from the sort of inside-out know-how one needed to create on an Apple II to the kind of perfunctory know-how one needed to navigate the surface of the Macintosh—one that amounted to the kind of knowledge needed to click this or that button. The Macintosh was democratic only in the manner any kitchen appliance was democratic.

Along with the way in which terms such as *transparency, customization,* and *user-friendly* were used, altered, and eventually turned inside out en route to the release of the Macintosh, Apple's redefinition of the overall philosophy of personal computing exemplified just one of many reversals that abounded in the ten-year period from the mid-1970s to the mid-1980s. In relation to the crucial change that took place in the mid-1980s from open, flexible, and extensible computing systems for creativity to ones that were closed, transparent, and task oriented, the way in which the Apple Macintosh was framed at the time of its release in January 1984 represented a near-complete purging of the philosophy promoted by Engelbart, Kay, and Papert. This purging of the recent past took place under the guise of Apple's version of the user-friendly, which among other things, pitted itself against the supposedly "cryptic," "arcane" "phosphorescent heap" that was the command-line interface, as well as, it was implied, any earlier incarnation of the GUI.[54]

It is important to note, however, that although the Macintosh philosophy purged much of what had come before it, it did in fact emerge from the momentum gathering in other parts of the computing industry, which in 1982 and 1983 were particularly concerned with defining standards for the computer interface. Up to this point, personal computers were remarkably different from each other. Commodore 64 computers, for example, came with both a Commodore key that gave the user access to an alternate character set and four programmable function keys that with the Shift key could each be programmed for two different functions. By contrast, Apple II computers came with two programmable function keys, and Apple III, IIc, and IIe computers came with open-Apple and closed-Apple keys that provided the user with applications shortcuts such as cut-and-paste and copy, in the same way that the contemporary Command key functions.

No doubt in response to the difficulties this variability posed to expanding the customer base for personal computers, *Byte* magazine ran a two-part series in October and November 1982 dedicated to the issue of industry standards by way of an introduction to a proposed uniform interface called the Human Applications Standard Computer Interface (HASCI). Asserting the importance of turning the computer into a "consumer product," author Chris Rutkowski declared that every computer ought to have a "standard, easy-to-use format" that "approaches one of *transparency*. The user is able to apply intellect directly to the task; the tool itself seems to disappear."[55] Of course a computer that is easy to use is entirely desirable. At this point, however, ease of use is framed in terms of the disappearance of the tool being used in the name of "transparency"—which then means users can efficiently accomplish their tasks with the help of a glossy surface that shields them from the depths of the computer, instead of the earlier notion of transparency, which refers to a user's ability to open up the hood of the computer in order to directly understand its inner workings.[56] In some ways, then, Rutkowski's proposed HASCI marked "the beginning of an era of consumer-oriented computers," with the emphasis no longer on learning or creativity but rather on, again, a computer that appealed to the widest possible swath of consumers, who wanted to use ready-made hardware and software mostly for the accomplishment of tasks and who most certainly did not want to tinker with expansion slots or programming.[57]

Throughout the following year, *Byte* continued to publish special issues on "easy software" and standards, as well as articles and editorials on the philosophy of the user interface, on how "windowing is the most natural way to express task con-currency," on the role of metaphor in "man–computer systems," and on various other GUIs or menu-based interfaces that never caught on, such as VisiOn and the Starburst User Interface. Thus, no doubt in a bid to finally produce a computer that real-ized these ideas and to appeal to consumers who were "driv-ers, not repairmen," Apple unveiled the Lisa in June 1983 for nearly $10,000 as a cheaper and more user-friendly version of the Xerox Alto/Star, which sold for $16,000 in 1981.[58] At least partly inspired by Larry Tesler's Xerox PARC 1979 demo of the Star to Steve Jobs, the Lisa—designed by Tesler himself, who moved to Apple a year later in 1980—used a one-button mouse, overlapping windows, pop-up menus, a clipboard, and a trash can. As Tesler was adamant to point out in the 1985 article "Leg-acy of the Lisa," it was "the first product to let you drag [icons] with the mouse, open them by double-clicking, and watch them zoom into overlapping windows."[59] The Lisa moved that much closer to the realization of the dream of transparency with, for example, its mode of double-clicking that attempted to nat-uralize the Star's text-based commands by no longer making the user actively choose "OPEN" and "CLOSE" and instead hav-ing them develop the quick, physical action of double-clicking that bypassed the intellect through physical habit. More, its staggering 2048K worth of software and three expansion slots firmly moved it in the direction of a ready-made, closed con-sumer product and definitively away from the Apple II, which when it was first released in 1977 came with 16K of code and, again, eight expansion slots.

Expansion slots symbolized the direction that computing was to take from the moment the Lisa was released to the re-lease of the Macintosh in January 1984 to the present day. Jeff Raskin, who originally began the Macintosh project in 1979, and Steve Jobs both believed that hardware expandability was one of the primary obstacles in the way of personal computing's broader consumer appeal.[60] In short, expansion slots made standardization impossible (partly because software writers

needed consistent underlying hardware to produce widely functioning products), whereas what Raskin and Jobs both sought
was a system that was an "identical, easy-to-use, low-cost appliance computer." At this point, customization was no longer
in the service of building, creating, or learning. It was, instead,
for using the computer as one would any home appliance, and
ideally this customization would be possible only through software that the user dropped into the computer via disk, just as
one would a piece of bread into a toaster. Predictably, the original plan for the Macintosh had it tightly sealed so that the user
was only free to use the peripherals on the outside of the machine. Although team member Burrell Smith managed to convince Jobs to allow him to add slots so that users could expand
the machine's RAM, according to Steven Levy, Macintosh own-
ers were still "sternly informed that only authorized dealers
should attempt to open the case. Those flouting this ban were
threatened with a potentially lethal electric shock."[61]

That Apple could successfully gloss over the aggressively
closed architecture of the Macintosh while marketing it as a
democratic computer "for the people" marked just one more
remarkable reversal from this period in the history of computing. As is clear in the advertisement in Figure 18, which came
out in *Newsweek* during the 1984 election cycle, the Macintosh
computer was routinely touted as embodying the principle of
democracy. While it was certainly more affordable than the
Lisa (in that it sold for the substantially lower price of $2,495),
its closed architecture and lack of flexibility could still easily
allow one to claim it represented a decidedly *undemocratic* turn
in personal computing.

Thus, 1984 became the year that Apple's philosophy of the
computer as appliance, encased in an aesthetically pleasing exterior, flowered into an ideology. We can partly see how their
ideology of the user-friendly came to fruition through their
marketing campaign, which included a series of magazine ads,
along with one of the most well-known TV commercials of the
late twentieth century. In the case of the commercial, Apple

FIGURE 18. *Two-page advertisement for the Apple Macintosh from the November/December 1984 issue of* Newsweek.

took full advantage of the powerful resonance still carried by George Orwell's dystopian post–World War II novel *1984* by re-assuring us in the final lines of the commercial, which aired on January 22, 1984, "On January 24th Apple Computer will introduce Macintosh. And you'll see why 1984 won't be like '1984.'"[62] Apple positioned Macintosh, then, as a tool for and of democracy while also pitting the Apple philosophy against a (nonexistent) other (perhaps communist, perhaps IBM or Big Blue) who was attempting to oppress us with an ideology of bland sameness. Apple's ideology "saved" us, then, from a vague and fictional, but no less threatening, Orwellian, night-marish ideology. As lines of robot-like people, all dressed in identical grey, shapeless clothing, march in the opening scene of the commercial, a narrator of this pre-Macintosh nightmare appears on a screen before them in something that appears to be a propaganda film (see Figure 19).

We hear, spoken fervently, "Today, we celebrate the first

FIGURE 19. *Screenshot of the commercial advertising the release of the Apple Macintosh (directed by Ridley Scott), which aired on January 22, 1984, during the third quarter of Super Bowl XVIII.*

glorious anniversary of the Information Purification Directives." And as Apple's hammer thrower then enters the scene, wearing bright-red shorts and pursued by soldiers, the narrator of the propaganda film continues:

> We have created for the first time in all history a garden of pure ideology, where each worker may bloom, secure from the pests of any contradictory true thoughts. Our Unification of Thoughts is more powerful a weapon than any fleet or army on earth. We are one people, with one will, one resolve, one cause. Our enemies shall talk themselves to death, and we will bury them with their own confusion.[63]

Just before the hammer is thrown at the film screen, causing a bright explosion that stuns the grey-clad viewers, the narrator finally declares, "We shall prevail!" But who exactly is the hammer-thrower-as-underdog fighting against? Who shall prevail—Apple or Big Brother? Who is warring against whom in this scenario and why? In the end all that mattered was that at this moment, just two days before the official release of the Macintosh, Apple had created a powerful narrative of its un-questionable, even natural superiority over other models of computing, a narrative that continues well into the twenty-first century. It was an ideology that of course masked itself as such and that was born out of the creation of and then opposition to a fictional, oppressive ideology from which we users/consumers needed to be saved.[64] In this context the fervor with which Mac-intosh team members believed in the rightness and goodness of their project is somewhat less surprising. They were quoted in *Esquire* earnestly declaring, "Very few of us were even thirty years old. . . . We all felt as though we had missed the civil rights movement. We had missed Vietnam. What we had was the Macintosh."[65]

We can see how this transformation from philosophy to ide-ology took place partly through their design bible from 1988, *Apple Human Interface Guidelines: The Apple Desktop Interface*, in which we learn, first, of the importance of an interface that is utterly consistent and familiar and that provides a believable environment via visual metaphors, such as the trash can icon or images of file folders, so that "people can perform their many tasks." We are told, "People are not trying to *use computers*—they're trying to get their jobs done."[66] Of course, use, not the accomplishment of tasks, is what makes creativity and learning on a computer possible. Second, we learn of the importance of an interface that makes commands visible for the user—and "visible" is yet another reversal, for here it means not the abil-ity to see and understand the underlying processes but rather that "the screen displays a representation of the 'world' that the computer creates for the user. On this screen is played out the full range of human–computer interactions."[67] In fact, what we see on-screen is not "the full range" of possible human–computer interactions but rather a predetermined set of interactions, designed to appear as though it is a full range of interactions, from which the user must choose. If the interface is indeed a threshold between user and computer, then what the Macintosh interface offered was an entirely simulated environment for the user with no access at all to the machine on the other side.[68]

Again, though, the "believable environment" offered by the Macintosh was so appealing, so seductive that it was nearly impossible to see its clear limitations. Even nonfiction accounts of the Macintosh by non-Apple employees could not help but endorse it in as breathless terms as those used by the Macintosh team members themselves. Steven Levy's *Insanely Great,* from 1994, is a document remarkable for an endorsement of this new model of personal computing as wholesale as that of any Macintosh advertisement or guidebook. Recalling his experience seeing a demonstration of a Macintosh in 1983, he writes:

> Until that moment, when one said a computer screen "lit up," some literary license was required. . . . But we were so accustomed to it that we hardly even thought to conceive otherwise. We simply hadn't seen the light. I saw it that day. . . . By the end of the demonstration, I began to understand that these were things a computer *should* do. There was a better way.[69]

The Macintosh was not simply one of several alternatives—it represented the unquestionably right way for computing. Even when he wrote the book in 1993, Levy still declared that each time he turned on his Macintosh, he was reminded "of the first light I saw in Cupertino, 1983": "It is exhilarating, like the first glimpse of green grass when entering a baseball stadium. I have essentially accessed another world, the place where my information lives. It is a world that one enters without thinking of it . . . an ephemeral territory perched on the lip of math and firmament."[70] But it is precisely the legacy of the unthinking, invisible nature of the so-called user-friendly Macintosh environment that has precluded using computers for creativity and learning and that continues in contemporary multitouch, gestural, and ubiquitous computing devices such as the iPad and the iPhone, whose interfaces are touted as utterly invisible and, therefore, whose inner workings are de facto as inaccessible as they are invisible. That said, once roughly fifteen years had passed since the release of the Macintosh, critiques of this model of frictionless, closed computing began to surface in activist digital media poetics.

3

Typewriter Concrete Poetry as Activist Media Poetics

Analog Hacktivism

The third archaeological cut I make into reading/writing interfaces is the era from the early 1960s to the mid-1970s in which poets, working heavily under the influence of Marshall McLuhan, sought to create (especially, so-called dirty) concrete poetry as a way to experiment with the limits and the possibilities of the typewriter. These poems—particularly, those by the two Canadian writers bpNichol and Steve McCaffery and the English Benedictine monk Dom Sylvester Houédard—often deliberately court the media noise of the typewriter as a way to draw attention to the typewriter as interface. Further, since these poems are about their making by way of a particular writing medium as much as they are about their reading/viewing, if we read these concrete poems in relation to Marshall McLuhan's unique pairing of literary studies with media studies—a pairing that is also his unique contribution to media archaeology *avant la lettre*—we can reimagine formally experimental poetry and poetics as engaged with media studies and even with hacking reading/writing interfaces. Another key point of this chapter is that we can also reimagine McLuhan's work as equally influenced by concrete poetry, and so it too is an instance of media poetics—even *activist* media poetics.

The Poetics of a McLuhanesque Media Archaeology

As Siegfried Zielinski writes in *Deep Time of the Media*: "Do not seek the old in the new, but find something new in the old. If we are lucky and find it, we shall have to say goodbye to much that is familiar in a variety of respects."[1] At the heart of media archaeology is an ongoing struggle to keep itself from ossifying into a set of inflexible methodologies, as well as the attempt to keep alive what Zielinski calls "variantology"—the discovery of "individual variations" in the use or abuse of media, especially those variations that defy the ever-increasing trend toward "standardization and uniformity among the competing electronic and digital technologies."[2] Pulled between a desire to renovate media studies and the necessity to keep such a renovation consistently flexible and even indefinable, much media archaeology–aligned writing is marked by the sort of unexpected reversals reflected in the quote by Zielinski. If we are to take seriously Zielinski's call to not "seek the old in the new, but find something new in the old," then in light of our newfound awareness of ways in which digital interfaces frame both reading and writing, the typewriter emerges as a profoundly influential *analog* reading/writing interface. Further, typewriter poetry broadly and dirty concrete poetry in particular are extremely effective in how they draw attention to the limits and the possibilities of the typewriter as interface. When Andrew Lloyd writes in the 1972 collection *Typewriter Poems* that "a typewriter is a poem . . . a poem is not a typewriter," he gestures to the ways in which poets enact a kind of media analysis of the typewriter via writing as they cleverly undo stereotypical assumptions about the typewriter itself. A poem written on a typewriter is not merely a series of words delivered via a mechanical writing device, and for that matter, neither is the typewriter merely a mechanical writing device.[3] Instead, these poems express and enact a poetics of the remarkably varied material specificities of the typewriter as a particular kind of mechanical writing interface that necessarily inflects both how and what one writes.

That said, in order not to use media archaeology as a productive framework but to actually *do* media archaeology by uncovering media-related phenomena such as the typewriter and dirty concrete poetry produced in the 1960s and 1970s, rather than drawing on a more recent figure such as Zielinski or the earlier and equally influential Friedrich Kittler, we instead ought to draw on Marshall McLuhan as a media archaeologist *avant la lettre* who is also finely attuned to the literary.[4] Further, we ought to use Zielinski's invocations to "find something new in the old" by focusing our efforts on McLuhan's writing to reinvigorate studies of him that have, for far too long, focused almost exclusively on only three catchphrases enshrined in *Understanding Media*: (1) the medium is the message; (2) media as the extensions of "man"; and (3) the global village.[5] To venture beyond this early collection of his writings, originally published in 1964, is to discover long-out-of-print books such as the 1967 *Verbi-Voco-Visual Explorations*—a book not more relevant than *Understanding Media* but that clearly reflects McLuhan's engagement with a poetics of media studies both in the sense of poetic writing and as a conceptual framework for thinking about media.[6] To return to McLuhan's less-canonical texts is to significantly broaden his legacy to include a thoroughgoing engagement with innovative poetry—especially, the concrete poetry that was being produced internationally at the time, along with the dirty concrete poetry that was being written prodigiously in pockets across Canada, including Toronto, where McLuhan lived for most of his life.[7]

There is no doubt that poets writing verbi-voco-visually themselves were as strongly influenced by McLuhan as he was by them. For example, *Verbi-Voco-Visual Explorations* was published by the self-proclaimed "intermedia poet" Dick Higgins through his Something Else Press in the same year his press published the first major anthology of concrete poetry, Emmett Williams's *An Anthology of Concrete Poetry*.[8] Invested as he was in poetry that situated itself between two or more inseparable media, Higgins's notion of intermedia was obviously saturated with McLuhan's notions of the new electric age and the global village. As he wrote in his "Statement on Intermedia" in 1966, the year before publishing the two volumes by McLuhan and Emmett Williams:

> Could it be that the central problem of the next ten years or so, for all artists in all possible forms, is going to be less the still further discovery of new media and intermedia, but of the new discovery of ways to use what we care about both appropriately and explicitly?[9]

Clearly, Higgins thought of *Verbi-Voco-Visual Explorations* and *An Anthology of Concrete Poetry* as two parts of the same conversation. The former features a multilinear page design and typography that is constantly changing to reflect shifts in McLuhan's content. Moreover, *Verbi-Voco-Visual Explorations* is a book version of a 1957 special issue of the journal *Explorations* edited by McLuhan and Edmund Carpenter (note that the use of "verbi-voco-visual" in *Explorations* anticipated the 1958 use of the term by the group of Brazilian concrete poets, the Noigandres, also the founders of concrete poetry). In fact, both McLuhan and the Noigandres appropriated James Joyce's 1939 penning of the phrase "verbi-voco-visual presentiment" in *Finnegan's Wake,* with the concrete poets transforming the term into "verbivocovisual" in their "Pilot Plan for Concrete Poetry." The Noigandres reinterpreted it (in part by removing the hyphens, possibly to emphasize the interdependence of the three separate elements) to mean "coincidence and simultaneity of verbal and nonverbal communication . . . a communication of forms, of a structure-content."[10] McLuhan's use of the term maintained its connection to both a simultaneous mode of communication that was not simply verbal and the specifically poetics-related emphasis on the interdependence of form (or structure, as the Noigandres put it) and content.[11]

Significant for this chapter is that McLuhan wanted to renovate "verbi-voco-visual" so that it resonated with media studies as much as it did with poetics—a newly inaugurated field of media studies that made possible the observation that the 1950s brought into being an "electronic age" defined by a "secondary orality" that permitted an "instant awareness of a total situation. Oral means 'total' primarily, 'spoken' accidentally."[12] While typewriters were not yet electric in the 1950s, they did exemplify for McLuhan this return to the oral. He writes in the section "Verbi-Voco-Visual," "The 'reeling and writhing' of Lewis Carroll is close to the action of pre-typewriter reading and writing. The staccato stutter of the typewriter on the other hand is really close to the stutter that is oral speech. The typewriter is part of our oral revolution today."[13] This pairing of the literary with a study of media is absent from nearly all writing that explicitly calls itself media archaeology, a pairing that is McLuhan's critical innovation.

Further, this pairing forces us to read the work of innovative poets as performing studies of the limits and the possibilities of certain writing media. In this light McLuhan's observation, "Stephen Spender once suggested that the reason there is no more avant-garde experiment in literature is that this role has been assumed by the new media of expression," was not so much radical as it was a straightforward statement of contemporary poetry's unique contribution to media studies: the experimentation with the limits and the possibilities of writing media, broadly, and writing interfaces, more specifically.[14] Thus, in the case of typewriter poetry from the 1960s and 1970s, it was not simply that poets happened to use a typewriter to achieve certain effects but that they foregrounded what had always been the case: "The artist senses at once the creative possibilities in new media even when they are alien to his own medium. . . . *The artist is the historian of the future because he uses the unnoticed possibilities of the present.*"[15] By 1970 McLuhan had explicitly aligned both poet and artist as future historian. As he writes in *Culture Is Our Business*: "Poets and artists live on frontiers. They have no feedback, only feedforward. They have no identities. They are probes."[16]

While he does not explicitly credit McLuhan, Zielinski clearly takes up the notion of artists as probes and further extends it such that their media-oriented probings of the past and the present-as-future are inherently *activist*. He writes:

> Few activists, however, take the more daring path of exploring certain points of the media system in such a way that throws established syntax into a state of agitation. This is poetic praxis in the strict sense that the magical realist Bruno Schultz of Poland understood it: "If art is only supposed to confirm what has been determined for as long as anyone can remember, then one doesn't need it. Its role is to be a probe that is let down into the unknown. The artist is a device that registers processes taking place in the depths where values are created."[17]

With poetic practice framed as one that "throws established syntax into a state of agitation," insofar as it gives an account of the normally invisible—the taken for granted that nonetheless defines what can be said—the asyntactical, nonrepresentational dirty concrete poetry is activist media poetry par excellence. It probes, or reads, the new in old or standard uses of media such as the typewriter—a probing that in part foreshadows poets' nonstandard use of the digital computer's command-line interface in the early 1980s.

Literary DIY and Concrete Poetry

Prior to McLuhan and the concrete poetry movement, writers such as Jack Kerouac, William Burroughs, and Charles Olson are frequently cited as examples of those "for whom typewriting seems to have provided for a large degree of commonality in their thinking and writing practices," with Olson's 1950 "Projective Verse" being, in particular, a canonical text on the poetics of the typewriter.[18] "Projective Verse" is, however, almost always read in relation to Olson's breath-based poetics, with readers taking particular note of the following declaration: "It is the advantage of the typewriter that, due to its rigidity and its space precisions, it can, for a poet, indicate exactly the breath, the pauses, the suspensions even of syllables, the juxtapositions even of parts of phrases, which he intends."[19] Instead, the connection between the breath and the poet typing stands in for the typewriter's larger contribution: the way in which it allows a turn away from "manuscript, press, the removal of verse from its producer and its reproducer" and toward, as Olson implies, a practice in which form and content, medium and message, process and product are necessarily intertwined.[20] In fact, Olson's (typewritten) writing on the typewriter also both expresses and prefigures the movement in the 1960s and 1970s to democratize the process of writing poetry through writing *and* distribution that draw attention to the literary artifact as both an object created and mediated by the typewriter—techniques that essentially turn the artifact inside out.[21]

As I discuss in greater detail in the proceeding sections, this philosophy of making—especially exemplified by the typewritten, dirty concrete poem—erodes the division between surface and depth, inside and outside. Take, for instance, the statements of poetics that appear in the first anthology of concrete poetry, Emmett Williams's *An Anthology of Concrete Poetry* from 1967, and the now-canonical collection of concrete poetry *Concrete Poetry: A World View*, edited by Mary Ellen Solt, from 1968. While only about a third of the poems in each volume are obviously made with a typewriter, it's noteworthy that nearly all of the poems depend on the typewriter-inspired structure of the grid. Since all manual typewriters use monospace fonts—or fonts whose individual letters take up the same amount of space on the page, as manual typewriters can move only the same distance forward for each letter—these letters naturally form lines and columns. More, those poems that *are* typewritten are accompanied not with the usual statements about the author's intent with regard to the content or semantic meaning of the poem but rather with author statements that take the form of descriptions about the type of typewriter, typing techniques, and even the size of the paper used. Aram Saroyan declares:

> I write on a typewriter, almost never in hand . . . and my machine—an obsolete red-top Royal Portable—is the biggest influence on my work. This red hood hold [*sic*] the mood, keeps my eye happy. The type-face is a standard pica; if it were another style I'd write (subtly) different poems. And when a ribbon gets dull my poems I'm sure change.[22]

Dom Sylvester Houédard's statement describing his "type-stracts" (abstract, typewritten visual poems) is not only re-markably detailed but also notably focused exclusively on the writing process and the writing medium used (including his use and misuse of the typewriter) rather than on describing his intentions with regard to the final written product:

> my own typestracts (so named by edwin morgan) are all produced on a portable olivetti lettera 22 (olivetti himself/ themselves show sofar a total non interest in this fact) there are 86 typeunits available on my machine for use w/2-colour or no ribbon
>
>   - or with carbons of various colours—the maximum size surface w/out folding is abt 10″ diagonal—the ribbons may be of various ages—several ribbons may be used on a single typestract—inked-ribbon & manifold (carbon) can be combined on same typestract—pressures may be varied— overprints & semioverprints (1/2 back or 1/2 forward) are available—stencils may be cut & masks used—precise placing of the typestract units is possible thru spacebar & ratcheted-roller—or roller may be disengaged.[23]

This quote not only reads much like a do-it-yourself guide to writing typestracts but also—with individual letters rather than words as "units" and the page as a "surface"—aligns the DIY philosophy with a poetics that seeks to spur readers/

writers to move away from a poetry that is a delivery mechanism for semantic meaning and toward a poetry whose meaning is more about a process of making that takes place outside a cycle of consuming (through traditional reading practices) the already created. Loosely speaking, it is an open-source poetics that lays bare its mechanisms of creation.

We need look only at Houédard's untitled typestract in the Williams anthology to witness a self-conscious use of the typewriter-as-writing-medium as a way to create a process-oriented text rife with nonsemantic meaning (see Figure 20).[24] Here, Houédard takes full advantage of the typewriter's monospaced fonts to create interconnected parts that are gridlike, permutational, and pictorial and that—in the case of the square spiral that spells "atom"—only occasionally make semantic sense. In fact, with the exception of "atom," almost all of the visual structure of the poem is built from the letters comprising "atom," thereby atomizing the word.[25]

This shift to a DIY process of writing, which inevitably involves a fine-tuned attention to the particularities of a given writing medium, is echoed in Mary Ellen Solt's description of Ilse and Pierre Garnier's "poème mechanique," which "amounts to 'a transformation of work' to work-activity of the linguistic materials."[26] Ronald Johnson even more overtly thinks of his typewriter works along these same lines:

> As I am unable to think except on the typewriter, my poems have been, from the beginning, all 8½" x 11". This is not only misunderstood by the printers, it is ignored. And if one should happen to bring it to their attention they say—do it yourself. So I have. I have begun to make my own letters and to think in ink.[27]

The poems by Johnson that appear in the Solt collection further demonstrate the extent of the typewriter's influence, as these works go beyond a straightforward use of the writing medium by building on some of the essential capabilities offered

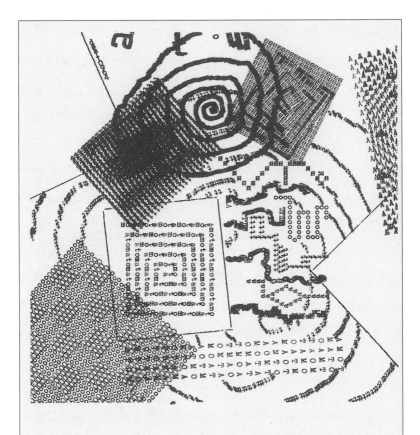

Dom Sylvester Houédard (1964)

"my own typestracts (so named by edwin morgan) are all produced on a portable olivetti lettera 22 (olivetti himself/themselves show sofar a total non interest in this fact) there are 86 typeunits available on my machine for use w/2-colour or no ribbon—or with carbons of various colours—the maximum size surface w/out folding is abt 10" diagonal—the ribbons may be of various ages—several ribbons may be used on a single typestract—inked-ribbon & manifold (carbon) can be combined on same typestract—pressures may be varied—overprints & semioverprints (½ back or ½ forward) are available—stencils may be cut & masks used—precise placing of the typestract units is possible thru spacebar & ratcheted-roller—or roller may be disengaged." (D.S.H.)

FIGURE 20. *Dom Sylvester Houédard's untitled typestract that appears in Emmett Williams's* Anthology of Concrete Poetry *from 1967. Copyright the Estate of Emmett Williams.*

FIGURE 21. *Ronald Johnson's "MAZE/MANE/WANE," which appears in Mary Ellen Solt's collection* Concrete Poetry: A World View. *Copyright 1968 by Ronald Johnson. Reprinted by permission of the Literary Estate of Ronald Johnson.*

by the typewriter (see Figure 21). Inspired by the grid space made possible by the monospaced font, we get a solid three-by-four block of intertwined (presumably hand-drawn) letters. Inspired by the ability to treat the page as a surface, we get a text whose negative space plays a role as fundamental as that of the positive space of the letters. Finally, from the ability, as

Houédard puts it, to "disengage" the roller and turn the page in any direction, we get an exploration of the typographical and visual connections between *Z* and *N, M,* and *W*.[28]

Further, in this same collection of concrete poetry edited by Solt, Dick Higgins—whose connection to McLuhan I touch on in the following section—explicitly makes the point not only that these concrete poems ought to be viewed rather than read (again, no doubt in part because of the absence of semantic meaning made possible by the way in which the typewriter transforms the page into a surface) but that a new mode of poetry is *required* because of new media of expression.

> As McLuhan says, you can't make the new medium do the old job. The information in a new poem can't be the same as the information in an old poem. . . . What interests me now is that new poetry isn't going to be poetry for reading. It's going to be for looking at. . . . I mean book, print culture, is finished.[29]

The foregoing clearly demonstrates a McLuhan-inspired mode of reading (writing) media such as the typewriter through concrete poetry that is also driven by a DIY sensibility.

From Clean to Dirty Concrete

Again, once we bring a media studies approach to bear on concrete poetry, we are immediately confronted with the fact that such an approach—in this case vis-à-vis the work of McLuhan—reveals that concrete poetry, no doubt along with the whole lineage of visual writing, is obviously *media* poetry. Further, we find that concrete poetry is not a homogenous field of writing but rather one that encompasses an extraordinary range of poems whose meaning is entirely tied to an equally extraordinary range of writing media used to create visual effects. While it may be true, as Marjorie Perloff points out, that when concrete poetry first emerged in the 1950s and the 1960s, it "presented itself as a coherent international movement with clearcut theories and practices," by the end of the 1960s and throughout the 1970s concrete poetry practices varied so widely that works from this era "showed anagrammatic dispersion and affirmed only the de-centring of all systems, the rejection of truth, origin, nostalgia, and guilt."[30]

In the face of canonical poems from the first generation, such as those by Eugen Gomringer or Augusto and Haroldo de Campos, which appear to be either made (or to be more precise, in those cases where manuscript versions are not available, the *published versions* appear to be made) with dry transfer lettering or typeset in lead—as well as all those poems created with stencils, stamps, xerography, letterpress, and so on—those concrete poems from both the first and the second generation of the movement whose form and content are inextricable from the typewriters used to create them stand apart as a (sub)genre. Further, those works from the second generation that might be called "dirty concrete" are even more obviously unique in that they use the typewriter—often in conjunction with the mimeograph machine—to push the limits of readability and interpretation by taking Robert Creeley's dictum, popularized by Olson's "Projective Verse," "form is never more than an extension of content," and generally turning form *into* content.[31]

The term "dirty concrete" is widely enough known that critics such as Marjorie Perloff can, in a discussion comparing Gomringer's and Steve McCaffery's work, simply mention in passing the distinction between clean and dirty concrete poetry without worrying about a readership who might not be familiar with the term.[32] That said, the term is commonly used to describe a deliberate attempt to move away from the clean lines and graphically neutral appearance of the concrete poetry from the 1950s and 1960s by Gomringer in Switzerland, the Noigandres in Brazil, and Ian Hamilton Finlay in England. Such cleanliness was thought to indicate a lack of political engagement broadly speaking and, more specifically, a lack of political engagement with language and representation. As renowned French poet Henri Chopin wrote in 1969, a year after the failed worker/student protests in France:

> 1968 was the year when man really appeared. Man who is the streets, HIS PROPERTY, for he alone makes it. . . . Yes, 1968 saw this. And for all these reasons, I was, and am opposed to concrete poetry, which makes nothing concrete, because it is not active. It has never been in the streets, it has never known how to fight to save man's conquests: the street which belongs to us, to carry the word elsewhere than the printing press. In fact, concrete poetry has remained an intellectual matter. A pity.[33]

Perhaps in response to criticisms such as those by Chopin that accused concrete poetry of being overly intellectual and not nearly concrete enough in the sense of being in and with the world, it was exactly around 1968 or 1969 that clusters of poets (mostly in Canada and the United Kingdom) began to produce concrete poems that deliberately courted a visual and linguistic nonlinearity and illegibility by putting the typewriter to the test. As these poets created smeared letters with inked ribbons or different carbons while turning and twisting the page, the result was often the imprint of letters that appeared literally dirty or rough around their edges.

Despite the references and the discussion around "dirty concrete," there is no clear written account of who first used this term. Tracking the evolution of the term and its shifting valences is instructive insofar as it points toward a kind of activism through a particular mode of typewriter poetry.[34] That is, it seems "dirty concrete" was used in loose conversation in the relatively small community of experimental Canadian poets and critics throughout the late 1960s and early 1970s as a viable, more politically activist alternative to clean concrete.[35] Still, it is worth noting that the earliest *written* record of the term appears in a letter from Stephen Scobie to bpNichol in 1968, in which there is a strong suggestion that Nichol came up with the term. Describing his own work to Nichol, Scobie writes, "You'll notice the difference in my work & a lot of the stuff you publish in Gronk—it is I think what you called the difference between 'clean' and 'dirty' concrete. . . . But the Canadians, especially bissett of course, are dirty. You mix the two, but I sense you're more at home in the dirty stuff." While this seems to suggest Nichol was the originator of the term, in another letter from Scobie to Nichol in 1971 there is the suggestion that Scobie and not Nichol came up with the term. Writes Scobie, "I never meant 'dirty' to be a term of disapprobation. I've got nothing against it: clean/dirty, like expressionist/constructivist, was always intended as a (possibly helpful, possibly not) description."[36]

Another written definition appears in a 1970 letter Nichol wrote to Nicholas Zurbrugg, the editor of *Stereo Headphones*, for a special issue Zurbrugg was working on called "The Death of Concrete," which includes the previous statement from Chopin and reinforces the fact that by the late 1960s and early 1970s there was a sense among concrete poets that the movement was stagnating around the clean form that had dominated the concrete poetry anthologies edited by Williams and Solt.[37] Here, Nichol echoes Chopin's concerns that clean concrete had become overly intellectual:

> concrete can become as big a trap as anything unless one
> stays open and flexible and is willing to keep seeking new
> exits and entrances with regard to the poem. which is to say
> the limitations with con lie within the men practising it, or
> within, say, a particular definition of it. . . . Stephen Scobie
> wrote to me from Vancouver and talked about the difference
> between 'clean' and 'dirty concrete.' by that definition we
> were all dirty. bruitist i suppose. for too many people con-
> crete is a head trip, which is to say, an intellectual trip. . . .

for most people i know it's a gut experience. i suppose you
could say that the 'concrete' in 'concrete poetry' has cracked
up but it sure as hell ain't dead.[38]

The term was likely then put into broader circulation by way of
bill bissett's 1973 "a pome in praise of all quebec bombers" in
pass th food release th spirit book, which as Jack David describes
it, "begins with the phrase 'dirty concrete poet' repeated twice,
then changes to 'the concrete is dirty dirty,' 'sum like it clean
what dew they ooo.' . . . The comparison presents the clean or-
dered life of a capitalist system and the dirty chaotic life of the
lower classes" (see Figure 22).[39] David's reading is nicely reinforced by other poems in the collection, such as "nefertiti," in
which bissett writes, "this aint no capitalist / pome ium tirud
uv finding / th ownr aint in arint yu."[40] Here, bissett aligns clear
semantic meaning and transparent, representational language
that the reader ought to passively consume with capitalism,
and consequently, he rejects both by way of nonstandard and
inconsistent spelling, syntax, spacing, and visual appearance
(e.g., overwriting with both the typewriter and the cracked im-
perfections of dry transfer lettering).[41]

That said, many of the poems in this collection push so hard
against semantic meaning in service of the nonstandard and
that which cannot easily be consumed that the results are often
not pictures or poems of or about anything so much as they are
inventive geometric designs that take advantage of the capabilities of the typewriter. Take, for example, the three gridlike
poems in the collection, all made with a different typewriter
than that used in "a pome in praise of all quebec bombers" (in
fact, it appears bissett used at least three different typewriters
in the creation of *pass th food release th spirit book*). These poems
all overlay the letter *Q* with horizontally and vertically aligned
o's and *n*'s to create dense, abstract designs (that are more or
less of a phallus).

In an email correspondence to me, bissett reinforced my
sense that the typewriter as a writing medium played a crucial

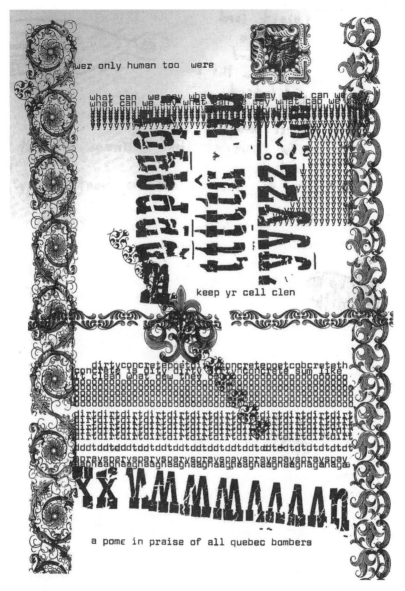

FIGURE 22. *bill bissett's "a pome in praise of all quebec bombers," which appears in his 1973* pass th food release th spirit book. *Reprinted by permission of Talonbooks.*

role particularly in the creation of concrete poems that explored the page as a surface and letters as units of composition: "ibm elektrik was a big fave n smith corona elektrik as well yu cud repeet lettrs i usd olympia as well i reelee enjoyd leeving my fingr on a lettr n feeling it type a whol long line take my fingr off th lettr key n it stoppd that was at leest th smith corona was xcellent 4 th tapestree kind uv konkreet."[42] By "tapestree" poetry, bissett likely means texts, such as "a pome in praise of all quebec bombers," whose designs and/or pictorial elements make them akin to a textile that is meant to be viewed rather than read. Unlike a fabric-based tapestry, however, these tapestry texts feature the typewriter not only as a writing machine capable of creating visual patterns but as one whose visual patterns, like a sewing pattern, lay bare the means for their own creation.

bpNichol, Dom Sylvester Houédard, Steve McCaffery

Steve McCaffery wrote presciently in 1986: "McLuhan saw the fundamental strength of technology as neither instrumental nor destructive but rather as rhetorical. Technology persuades towards modification and change; it is ideological software whose implications are both pre and post political. Technology does not serve so much as modify; it simultaneously promises and threatens change.[43] This section focuses on the work of two Torontonian contemporaries of the Vancouver-based bissett, both of whom were influenced by and even lived near Marshall McLuhan—bpNichol and his typewriter poetry and Steve McCaffery and his thoroughly dirty concrete poem Carnival— and that of English poet Dom Sylvester Houédard, who was one of the most prolific typestract poet-artists and who exerted significant influence over both Nichol and McCaffery.

While Nichol, who collaborated extensively with bissett through the mid- to late-1960s, rarely wrote concrete poems as visually dirty as McCaffery's, his writing career was defined by a McLuhan-inflected desire he expressed in his 1966 "Statement":

now that we have reached the point where people have finally come to see that language means communication and
that communication does not just mean language, we have
come up against the problem, the actual fact, of diversifi-
cation, of finding as many exits as possible from the self
(language/communication exits) in order to form as many
entrances as possible for the other.[44]

If according to McLuhan's famous 1964 declaration the "the medium is the message" and even the electric light is a communication medium that is in itself a message, then Nichol believed
he could realize his desire to move beyond an ego-based poetry
of self-expression through concrete poetry experiments with
diverse writing media generally and, at least until about 1973,
the typewriter in particular.[45] Also influenced by McLuhan,
McCaffery in *Carnival,* more than most dirty concrete poets
from this era, pushed the typewriter machinery to its limits
while also pushing his writing to the limits of legibility, inter-
pretability, and readability. This section delves into not only
how—especially in the context of the political turmoil of the
1960s and 1970s—both poets' probings of the limits and the
possibilities of communication via the typewriter as writing machine anticipated the *digital* media probings of the e-literature
authors I discuss in chapter 1 but also how they were activist in
the terms I outline in the previous section. Their typewriter/
dirty concrete poetry represented a push to unsettle what Zielinski calls "established syntax" by exploring and even hacking
the typewriter as a media system. As McCaffery himself wrote
of *Carnival* in 1975, this work "developed into an *exploration of
technologic tension*—that's to say how far you can push and extend the capabilities of textual-textural mechanics."[46]

Nichol's and McCaffery's work quite literally shows us
how media determine what can and cannot be said. Further,
using a medium such as the typewriter in ways not intended
or endorsed by the manufacturer results in the creation of
new modes of communication and, thus, newly communicated

content. These typewriter/dirty concrete poems are thoroughly activist media poems.[47] They are even activist in the sense that McLuhan was imagining in 1966 when he wrote in *Astronauts of Inner-Space: A Collection of Avant-Garde Activity*, notably alongside contributions by first-generation concrete poet Decio Pignatari and second-generation typestract poets Dom Sylvester Houédard and Franz Mon: "If politics is the art of the possible, its scope must now, in the electric age, include the shaping and programming of the entire sensory environment as a luminous work of art."[48] Politics as art; art as politics.

bpNichol's first substantial collection of typewriter concrete, *Konfessions of an Elizabethan Fan Dancer*, was not published in Canada until 1973 (it was first published in 1967 by British concrete poet Bob Cobbing's Writers Forum Quartos). By that time, Nichol had started to use the typewriter only as a means of transcribing handwritten poems. As he put it in an interview with Raoul Duguay, handwriting provided him with a more "intimate involvement with the architecture of the single letter," whereas the typewriter—for all that it offered—was limited to reproducing identical letters with a similarly identical, repetitive striking on the keys.[49] Still, the mechanism of the typewriter was a foundational element of Nichol's poetics, as his media experiments from 1965 to 1967 inaugurated a lifelong, acute attention to the contours of an astonishing range of writing media that extended to the pen and pencil as much as to the mimeograph stencil, letraset, photoduplication, the rubber stamp, embossing, and die-cutting.[50]

The poems in this collection are fairly representative of Nichol's typewriter poems in that they depend on this writing machine to produce anything from a sound poetry score ("Cycle #22") to combinatorial experiments ("Cycle #23"), figurative poems ("Easter Poem"), abstract poems, and often some combination of all four modes of concrete. In terms of the more abstract poems (which are, I believe, not nearly abstract or dirty enough to call "typestracts"), whereas bissett and McCaffery often court the abstract via typewriter as a way to disrupt a

clean, transparent writing/surface, Nichol's abstract typewriter poems—even when they are clearly in dialogue with bissett and McCaffery—are notably clean and legible even while
they too strive to disrupt transparency. For example, "The Return of the Repressed" picks up bissett's experiments with the
visual connections between the monospaced *O* and *Q*, and instead of overlaying the two letters to create noise, Nichol creates a visual pun as a way to enact the psychological life of letters (see Figure 23). Not only is "The Return of the Repressed"
resolutely for looking at rather than through, but our appreciation of what we are looking at is entirely dependent on our
how-to knowledge—how we write, how we write letters, how
we write with typewriters, how the typewriter works. Further,
it is this emphasis on the active how-to that makes the poem
(however modestly) *activist.*

Published several years before *Konfessions of an Elizabethan
Fan Dancer* and during his most intensely productive years with

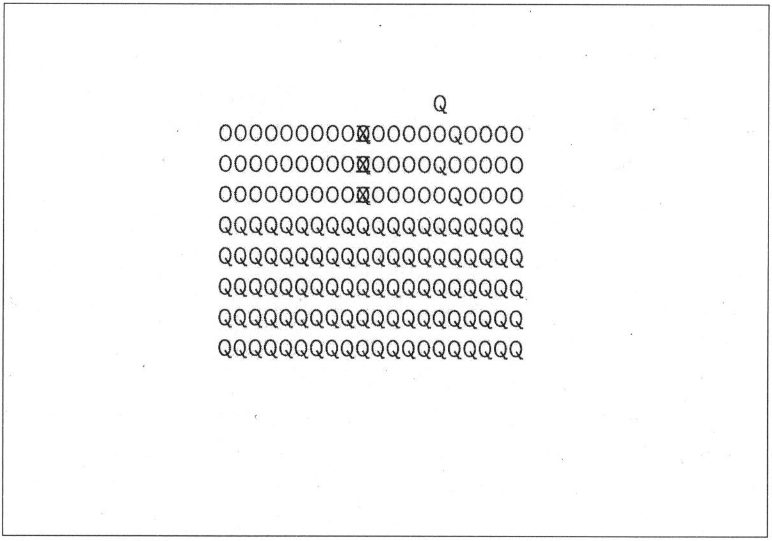

FIGURE 23. *bpNichol's "Return of the Repressed," which appears in his
1973 collection* Konfessions of an Elizabethan Fan Dancer. *Reprinted by
permission of Eleanor Nichol on behalf of the bpNichol Estate.*

the typewriter, his 1968 "The Complete Works" is Nichol's ultimate homage to this machine that is both a visual and a permutational writing device (see Figure 24).[51] Using the typewriter,
he attempts to reproduce the precise order/appearance of the
letters and characters on the QWERTY keyboard. Ultimately,
however, the poem is only an orderly reproduction of the typewriter key *functions* and not a reproduction of the appearance
of the keyboard, as the Shift key on typewriters from the 1950s
and 1960s supplanted the need to include lowercase letters on
the keyboard. This reading of the poem as a how-to guide to
writing—a poem whose meaning and content are entirely dependent on the typewriter—is reinforced by the fact that Nichol's poem is preceded by Aram Saroyan's 1965 "The Collected
Works."[52] Conceptually, of course, the two poems are nearly
identical. However, not only is Saroyan's poem written from a
very different typewriter (as it excludes mathematical symbols
and the numeral 1 and includes a key for ¾), but the published
version is typeset and so lacks the media-specific focus of Nichol's version. As such, Nichol's "The Complete Works" becomes
even more a poem about the potential of potential (reinforced
with an asterisk reminding us of "any possible permutation of
all listed elements") that is here necessarily tied to the material differences and particularities of the typewriter. The poem
not only is a DIY guide to poem writing in that it consists of
eighty-eight different combinations of eighty-eight letters and
characters but also is an example of potential literature even
more brilliant than Raymond Queneau's resolutely readable but
infamously never completable *Cent Mille Milliards de Poèmes,* as
Nichol's poem is both utterly unreadable and uncompletable in
its concern with the writing process as product. As he put it in
1976, once again using McLuhanesque terms that recall the dictum from *Understanding Media* that "media are the extensions
of man," "When I was into typewriter concrete, it was as much
that I was into the typewriter as a *tool of the writer—as an extension of the writer.*"[53]

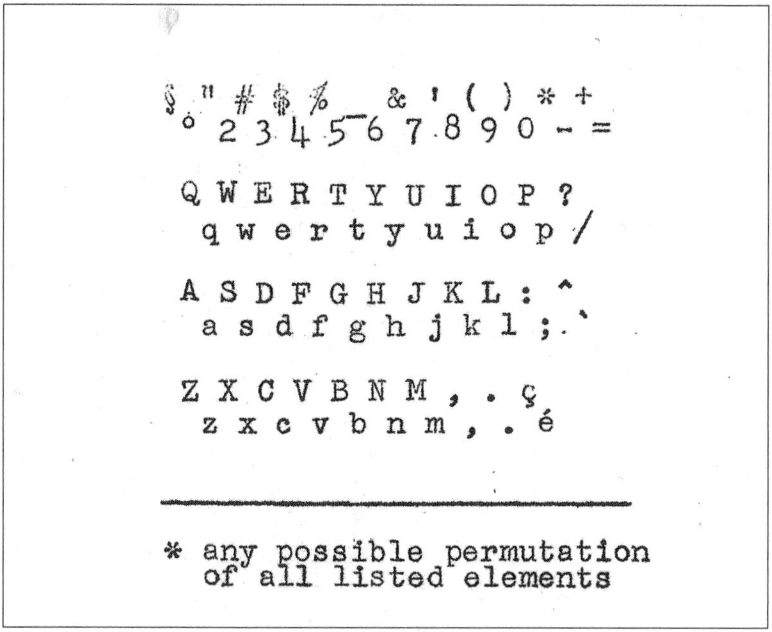

```
§  "  #  $  %     &   '  (   )   *  +
 °  2  3  4  5  6  7  8  9  0  -  =

Q  W  E  R  T  Y  U  I  O  P  ?
q  w  e  r  t  y  u  i  o  p  /

A  S  D  F  G  H  J  K  L  :  ^
a  s  d  f  g  h  j  k  l  ;  '

Z  X  C  V  B  N  M  ,  .  ç
z  x  c  v  b  n  m  ,  .  é
```

```
*  any possible permutation
   of all listed elements
```

FIGURE 24. *bpNichol's 1969 concrete poem "The Complete Works." Reprinted by permission of Eleanor Nichol on behalf of the bpNichol Estate.*

Though more readable and so less resolutely oriented toward writing that exemplifies process as product, Nichol's later collection of mostly typewritten poems *Translating Translating Apollinaire* (TTA)—a series of homophonic translations written between 1975 and 1979 of his own 1964 "Translating Apollinaire"—similarly explores the ways in which the typewriter engenders both visual and permutational writing. In its relentless exploration of different procedures for English-to-English translations, it has become something of a cult serial poem in certain experimental writing circles, spawning iterations such as Stuart Pid's *Translating translating translating Apollinaire* and Andrew Russ's *Translating, translating, translating Apollinaire, or, Translating, translating bp Nichol* (both from 1991). Nichol writes by way of an introduction:

May 27th 1975 en route from London England to Toronto
with Gerry Gilbert . . . in a mood of dissatisfaction re certain
aspects of my writing (always the feeling there is more one
should be learning—more limitations one should be pushing against & breaking down) i began this present series. In
my mind was the idea of a pure bit of research one in which
the creativity would be entirely at the level of the research,
of formal inventiveness, and not at the level of content per
se i.e. i recalled the first poem i had ever had published—
Translating Apollinaire in bill bissett's *BLEW OINTMENT*
magazine circa 1964 . . . & decided to put that poem thru as
many translation/ transformation processes as i & other
people could think of.[54]

Among the series of TTAs is a little-known collection called
*Sharp Facts: Some Selections from Translating Translating Apolli-
naire 26*—a series of TTA poems that are both typewriter and
photocopier poems. Given his love of the pun, a love he shared
with McLuhan, one of his favorite photocopiers was, not surprisingly, the Sharpfax copier machine (as they were called at
the time). Writing as an experienced writing-media technician
two years before his short McLuhan tribute, which I discuss
later, Nichol declares in the introduction:

> The translative system involved here entails the use of . . .
> copying machine disintegrative tendencies. Which is to say
> that an image fed through a copying machine over & over
> again (feeding the image of the image, & then the image of
> the image of the image, & so on) thru a great many genera-
> tions, disintegrates. & it does this differently depending on
> which type of copying machine you're using.[55]

The primary example of his copier machine poetics was created
on a Sharpfax—a series of poems that are by far the dirtiest,
most illegible of Nichol's works (see Figure 25). These poems
are dirty copier concrete (even though the original, thoroughly

legible poem he copied was made on a typewriter) whose content is the noise of media transmitting this same content. Nichol concludes his introduction by writing, "The ultimate goal of TTA 26 is to produce generational disintegrations on all the different types of copying machines. The analogue is one of a transmission thru time, a speeding up of the break-down process given information in a purely machine context. *In this case the machine is the message. The text itself ultimately disappears.*"[56] These are poems of and about writing media—poems that are not interested in their own illegibility *per se* so much as they are invested in reading, vis-à-vis writing, the typewriter through the copier machine.

With the abundance of near-McLuhanisms throughout Nichol's critical writing, the media theorist's influence on Nichol appears to be broadly conceptual in that he seems to have both internalized many of McLuhan's key precepts and admired McLuhan as a writer of *poetics*. Within a year or so after McLuhan's death in 1980, Nichol wrote "The Medium Was the Message," in which he made clear that although they did meet, McLuhan's influence "was model, a style or way of thinking":

> He understood that writing was not simply what is written but rather, in the very way you approached it, the very terms you set for yourself, became and becomes a strategy for living, a model for how to deal with the "reality" of the world. He showed how the medium became the message and how the most profound thot becomes cliché, becomes archetype. He showed us, too, a way to re-energize the language, the word world.[57]

While McLuhan undoubtedly had an influence over Nichol's concrete poems, which doubled as investigations into writing media, Steve McCaffery arguably had an even stronger, if not complementary, influence. Around 1967 McCaffery moved to Toronto from England and inaugurated several decades of collaboration between the two. Formally, these collaborations

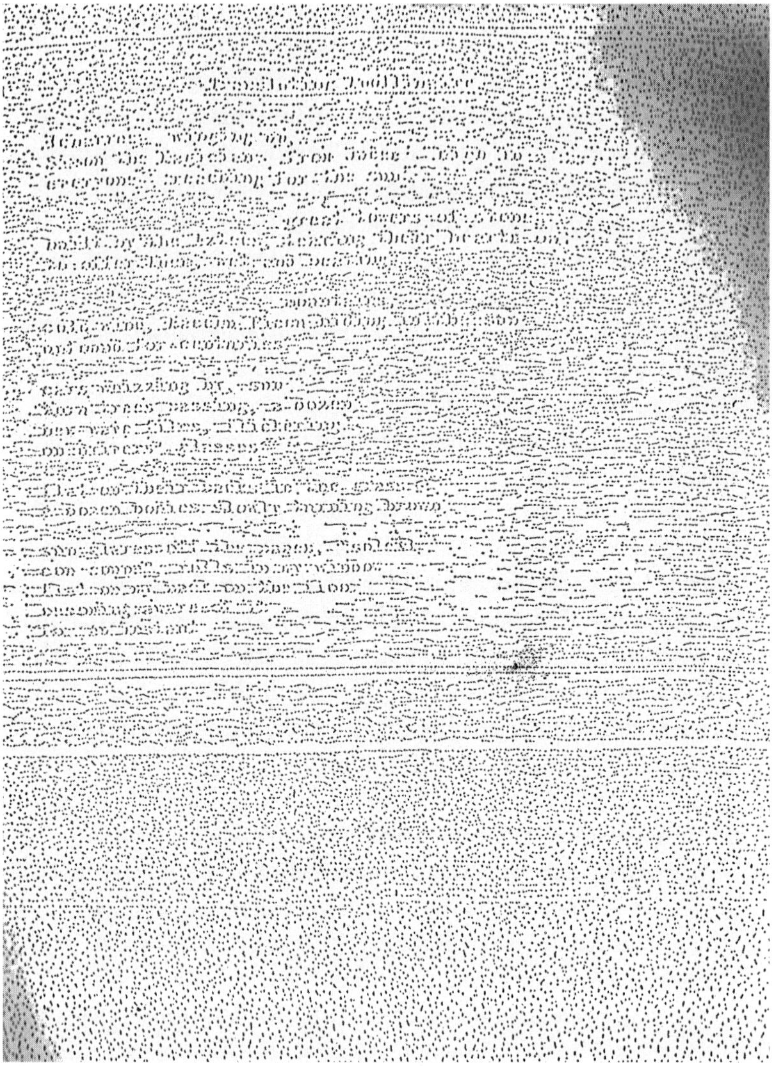

FIGURE 25. *A copier poem made on a Sharpfax copier machine that appears in bpNichol's 1980* Sharp Facts: Some Selections from Translating Translating Apollinaire 26. *Reprinted by permission of Eleanor Nichol on behalf of the bpNichol Estate.*

took place through their creation of the Toronto Research Group (TRG) and, along with Paul Dutton and Rafael Barreto-Rivera, through their sound poetry collaborative the Four Horsemen.[58] In a 1986 discussion with Geoff Hancock about the importance of McCaffery's arrival, Nichol declared:

> That made a huge difference in my life. Here was someone who was concerned with the same issues, and covered the same ground from his own angle for his own reasons. Steve and I are very dissimilar writers. But we share a lot of concerns. I always concerned myself with design, typeface, and papers on the press though you wouldn't necessarily know it by looking.[59]

Even as McCaffery's work throughout the late 1960s and the 1970s was more obviously concerned with moving as far away as possible from writing that attempted to represent reality through experiments that one could say were performative (in how they drew attention to the page as writing canvas and the mechanical means for producing letters that acted as the basic unit of composition), Nichol's work was equally concerned with the same issues. Further, judging from TRG documents Nichol and McCaffery both penned, the typewriter as an object of and a *crucial means for* thinking was a nexus for their shared interests.[60]

> We've always typed. We type with maybe one of us typing what's in our mind and then we kick an idea around. And then maybe I dictate to Steve while he types. And maybe I'm typing, and he's dictating to me. And I'm adding something as I think of it. And then we go over it, and go over it. So it happens at the time of writing.[61]

While the typewriter may have been a productive intermediary for both Nichol and McCaffery in that it acted as a kind of catalyst to improvisational thought, in McCaffery's

single-authored works such as *Carnival* it operated largely as
a means for his attempts to achieve a calculated annihilation
of semantic meaning.[62] Even though it is just one of many
typestracts McCaffery created during this period, *Carnival* is
his most well-known concrete poem and certainly his most
sustained work of typewriter poetry—it is even, arguably,
the most well-known work of any typewritten dirty concrete.
That said, to have a complete understanding of *Carnival* it is
important to note that McCaffery was clearly experimenting
with the limits of the genre in the years leading up to, during,
and even after *Carnival*. With 1969 as his most productive year
with the typewriter—during which time he created numerous
typestracts, among them "Broken Mandala," which reflected
his "desire to capture the force of sheer imprint . . . imprint as
it registers as a gestural, manual trace, a hand-print impressed
upon language"—McCaffery continually sought to find the limits of a literary-based material gesture.[63] While the pieces in the
"Broken Mandala" series do include the typewritten repetition
of one legible word, "from," the process of creating this textual
repetition is used to create utterly abstract, painterly shapes,
layers, and textures such that "from" enacts its semantic meaning of transformation on nearly every level. We can see how
such an engagement with the limits of the readable gesture via
typewriter finds its logical conclusion in works such as the 1975
"Punctuation Poem" and, ultimately, in 1980's "Suprematist Alphabet."[64] In fact, even as the former consists only of wavy rows
of commas, semicolons, and periods (such that it goes beyond
concrete poetry's established form of the grid), the latter pur-
sues the utter annihilation of semantic meaning and representationality through perfectly symmetrical lines of overwriting.
The result of these superimpositions is a text that is simultaneously clean and dirty—neatly typewritten letters of the alphabet become increasingly blurry and unrecognizable with each
equally neat overwriting.

In terms of the literary milieu in which *Carnival* was created,
while I am, in part, arguing in this chapter that there was a

uniquely Canadian context for the McLuhan-inflected dirty concrete poems of the 1960s and 1970s, which experimented with the material limits and the possibilities of the typewriter, given that concrete poetry was from the beginning a resolutely international movement it is worth pointing out that the progression of McCaffery's typewritten dirty concrete appears to have been strongly influenced not only by bissett and Nichol but also by the work of the English Benedictine monk Dom Sylvester Houédard.[65] At some point in the 1960s, McCaffery did briefly meet Houédard at Prinknash Abbey in Gloucestershire, but it is more likely that Houédard's influence acted largely through publications, which often included work by both poets, and perhaps indirectly through Houédard's correspondence with Nichol.[66] Houédard consistently published an astonishing range of typestracts from the early 1960s through the 1980s. While his work throughout the 1960s varied substantially in both technique and content—from the pictorial to the abstract (along the lines of the piece from the Emmett anthology I discuss in the previous section), the permutational, and the gridlike—the overall trajectory of his typewriter poetry was similar to McCaffery's. Houédard gradually focused exclusively on creating geometrically clean yet utterly abstract designs that may or may not be constructed with letters. For example, while his 1971 "earthbond" contains no text at all and whose shape is made entirely from typewritten hyphens and slashes, his 1975 "leaning on an angel" does contain the title of the text, yet the letters of the text itself are created with typewritten slashes, as is the "poem" itself—a series of lines constructed to form a blank-space ring around a lined circle in the center (see Figure 26).[67] The result is a tense interdependence between the semantic and the purely pictorial that is utterly *of* the typewriter.

As he tellingly wrote in 1979 of an early revelation he had about the possibilities of the typewriter that continued to play out through his writing career: "During 1945 I realised the typewriter's control of verticals and horizontals, balancing its mechanism for release from its own imposed grid, (and) offered

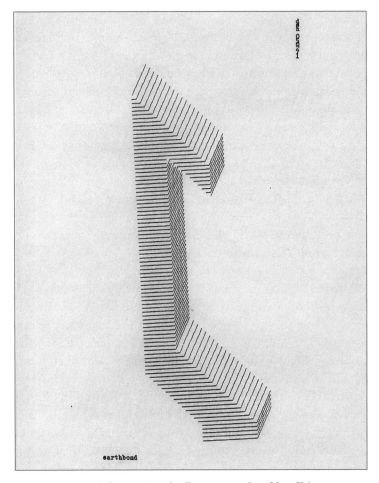

FIGURE 26. *Dom Sylvester Houédard's typestract "earthbond" from 1971. Sackner Archive of Concrete and Visual Poetry, by kind permission of the Prinknash Abbey Trustees.*

possibilities that suggested (I was in India at the time) the grading of Islamic calligraphy from cursive (naskhi) writing through cufic to the abstract formal arabesque, that 'wise modulation between being and not being.'"[68] Unquestionably, given the two aforementioned works, Houédard continued well beyond 1945 in his exploration of the movement between working within and without the constraints of the typewriter grid by way of both alphabetic and nonalphabetic characters.

The most important connection that Houédard offers to this study of Canadian dirty concrete is, however, his fascination with Marshall McLuhan. In the same year that McLuhan published *Understanding Media*—in which he makes the distinction between hot and cool media—Houédard wrote his 1964 "cool poem," whose shapes are created with the letters *c, o,* and *l,* though neither letters nor shapes clearly or obviously spell out "cool" (see Figure 27).[69] With hot media as that which is "low in participation" and cool media as that which demands high "participation or completion by the audience" (McLuhan lists the cartoon, the hieroglyph, and ideogrammic written characters as examples), Houédard's "cool poem" is surely more McLuhanesque than a 1960s homage to "cool" as the nifty or neat.[70]

More revealing, a 1965 letter from Houédard to bpNichol opens with a small typewriter poem in the form of a greeting and then proceeds:

> in canada dyou know marshall mcluhan?—his books
> have big influence on especially furnival—cor bougre—
> just looked up address—he is yr neighbour—29 wells hill
> toronto- 4—like photography liberated art from having to
> be a reporters lens—radio-tv-&c liberates poetry from (&
> prose from) i mean ALL communication artwise from being
> written descriptive report—so abstract or concrete poetry
> is cool in mcluhan sense.[71]

Even though he was in possession of McLuhan's mailing address, I have not yet found evidence to suggest that Houédard and McLuhan ever met or corresponded. Recalling that only a year later contributions by both appeared in *Astronauts of Inner-Space,* separated by a few pages, this letter serves, however, only to further confirm the tight interconnection between McLuhan's canonical and noncanonical media writings and typewriter/typestract poetry.

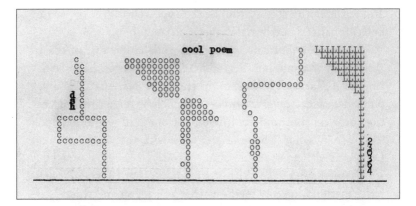

FIGURE 27. *Dom Sylvester Houédard's typestract "cool poem" from 1964. Sackner Archive of Concrete and Visual Poetry, by kind permission of the Prinknash Abbey Trustees.*

Given the mid-1960s literary milieu of Toronto and the active correspondence among concrete poets in Canada, Europe, and South America, it is not surprising that by 1967 McCaffery had begun work on his two-panel *Carnival*. Unlike Houédard, whose entire oeuvre was defined by his work within or around the stringency of the carriage, McCaffery created the first panel of *Carnival* from 1967 to 1970 with the typewriter and the addition of masks; the latter was his solution to the problem of how to work against the linearity of the typewritten line, which could bring (back) into play traditional reading and writing habits (see Figure 28). As he explains in the following quote, the mask is a physical interposition that excludes and fractures text—it moves the work outside the control of the author and his intentions in that its exclusions are determined by the material qualities of the mechanism of the mask itself:

> *Carnival* was essentially a cartographic project; a repudiation of linearity in writing and the search for an alternative syntax in 'mapping.' . . . The panels grew directly through the agency of the typewriter and through the agency of marginal link-ups. . . . As a mask bled off a page I would

devise another shape that picked up the bleed of the text at the margin. . . . The mask came about as a way to create a painterly shape by censoring the flow of typewritten line. . . . It's important to remember that the mask excludes and deletes much of the written text. What results are deliberately induced fragments, parts of inscription whose terminations and commencements are not determined by a writing subject or a logical intention but by a material, random intervention.[72]

The culmination of McCaffery's work with and against both mask and typewriter in the first panel of *Carnival* is a typestract that very nearly explodes visual and semantic representationality. I write "nearly" because, in addition to the masking, it encompasses a broad range of concrete poetry forms and techniques, including the concrete poem, whose form literalizes its content (take, for example, the section that repeats across the page "eyeleveleyelevel" at eye level) and so actually merges visual with semantic representationality instead of rejecting representationality altogether. Moreover, McCaffery pushes the physical possibilities of the page and the book to their limit not in using single pages with margins whose supposedly empty space is used to frame "meaningful" text but rather by writing over the edges of each of the sixteen 8.5-by-11-inch pages so that we see whatever blank space remains as meaningful in itself. These individual pages in turn are perforated and arranged in sequential book form, accompanied by the following instruction: "In order to destroy this book please tear each page carefully along the perforation."[73] Thus, the final form of the poem may not be a book whose pages proceed linearly but rather, only if the reader chooses to follow the instructions to destroy the book, a 44-by-36-inch square.

The second panel was created between 1970 and 1975— and incidentally, was later coedited for Coach House Press by bpNichol, to whom McCaffery dedicated this second panel (see Figure 29). Here, McCaffery extends his experiments with

FIGURE 28. *The sixth 8.5-by-11-inch sheet that constitutes the first panel (1967–70) of Steve McCaffery's* Carnival. *The image appears here in monochrome, but the panel was originally published as polychrome. Courtesy of Steve McCaffrey.*

writing media to include, in addition to the typewriter, "xerography, xerography within xerography (i.e. metaxerography and disintegrative seriality), electrostasis, rubber-stamp, tissue texts, hand-lettering and stencil."[74] Building on bissett's and Nichol's typewriter work in dirty mimeo, this second panel uses processes such as photocopying copies to significantly muddy or intervene in any kind of visual clarity. The result is, in part, a deliberate, activist-oriented courting of media noise that McCaffery describes in the introduction as a "structure of strategic counter-communication." It is one that reminds us of the material workings of writing media that, alongside the visual nature of the written word that in itself communicates, always already shape every contour of communication. As such, in comparison with the first panel, the second panel is generally "more" on nearly every level—McCaffery uses more writing media to push these media to their use limits. There are also more semantically clear, readable fragments of texts in this panel, even if these fragments exhort us to write rather than to read, to look at our words rather than to look through them: "you must write / upon it you must write / upon the page that there is / white upon the page." While greater readability might seem a counterintuitive development from the first to the second panel, fragments such as the foregoing remind us that despite the experiments of the first panel, the complete annihilation of semantic meaning is neither possible nor desirable. The point is, rather, to *activate* our sense of the profoundly materialist and the multidimensionality of the reading/writing process. (Understanding the sophistication of the second panel of the simultaneously viewable and readable *Carnival* is akin to understanding that when McLuhan asserts "the medium is the message" he is not asserting there is no such thing as a message at all—only that all messages are mediated and that all media bear with them a message.)

Finally, to turn to the ways in which the media archaeology approach underlying this chapter involves reading "old" media against "new" media—to disabuse us of this belief in

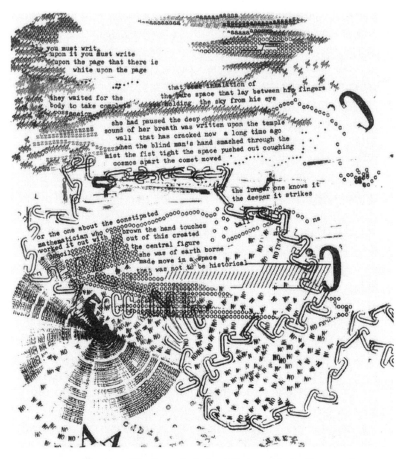

FIGURE 29. *The second 8.5-by-11-inch sheet that constitutes the second panel (1970–1975) of Steve McCaffery's* Carnival. *This second panel was originally published as polychrome. Courtesy of Steve McCaffery.*

the progression from new to old as much as to make visible the invisible aspects of contemporary media structures—what is also significant about McCaffery's project is that the typewritten text, the stamps, the various traces of writerly labor and the physical world (in the form of smudges or the slight bleed of ink) turn it into a work in which the surface *is* the depth and the making of the work *is* the meaning.[75] I contend this in

spite of McCaffery's own assertions in one of the first issues of
$L=A=N=G=U=A=G=E$ *Magazine* (which was republished as the
introduction to the second panel) that "Carnival is product and
machine, not process. . . . It must stand objective as a distancing
and isolating of the language experience."[76] He goes on to say,
"It is language presented as direct physical impact, constructed
as a peak, at first to stand on and look down from the privilege
of its distance onto language as something separate from you.
But Carnival is also a peak to descend from *into* language."[77]
This turn, this descent into language, transforms it into a product, a representation of "the language experience" that is about
its processual descent. Christian Bök and Darren Wershler
rightly declare that "the panels do not diagram a premeditated
mission of intention for a product to be formed but diagram a
'spontaneous emission' for a conduct to be performed . . . the
mapmaker recording a process for producing a process of recording."[78] It is partly the evidence of the sheer volume of labor
executed over a period of years, all of which took place within
the stringent confines of a typewriter carriage, that makes the
work less about what is written and more about how it provides
a record of the labor of writing that doubles as a kind of how-to
guide to writing.[79]

Moreover, this processual, labor-oriented aspect of the
work gives rise to one possible media archaeology reading of
McCaffery's dirty concrete/typestracts that aligns work such as
Carnival with the current digital DIY movement and, especially,
with the programming language appropriately named Processing. In the introduction to Casey Reas and Ben Fry's *Processing:
A Programming Handbook for Visual Designers and Artists,* Reas
writes about how his time at the Aesthetics and Computation
Group at MIT was transformed after working with computer
scientist and graphic designer John Maeda, who described his
philosophy as one made possible by "dirty hands."[80] Writing
for Harvard's business school blog, Maeda declared, "In the
last few decades, technology has encouraged our fascination
with perfection—whether it's six sigma manufacturing, the

zero-contaminant clean room, or in its simplest form, '2.0.' Given the new uncertainty in the world however, I can see that it is time to question this approach—of over-technologized, over-leveraged, over-advanced living. The next big thing? Dirty hands."[81] With Maeda endorsing an approach to education and even a lifestyle driven by doing, by physically working with tangible materials, it is easy to see how such a philosophy brought about a shift in Reas and Fry from being "a consumer of software to a producer."[82] As a result, they created Processing not only as a means to relate "software concepts to principles of visual form, motion, and interaction" but also as a means to "increase software literacy in the arts." Explaining what software literacy means, they include a telling quote from Alan Kay (whom I discuss in depth in chapter 2), known for his pioneering work on object-oriented programming and windows-style graphical user interface design at Xerox PARC: "The ability to 'read' a medium means you can access materials and tools created by others. The ability to 'write' in a medium means you can generate materials and tools for others. You must have both to be literate."[83]

Casey and Reas's solution to the need for software literacy in the arts has been, then, to create a programming language that results not in a WYSIWYG interface, by which you can, for example, click a button to create a circle or a square, but rather in a scaled-down, simplified, even transparent language that binds users to what they produce. The difference between Processing and a largely WYSIWYG-based program such as Flash—which was popular for a time in the creation of digital poems—has mostly to do with the degree of access to making that has been built into each. The other, less obvious, but no less significant difference is that Processing is entirely open source, whereas Flash is, of course, entirely proprietary. The latter results in digital artwork, such as that by Young-Hae Chang Heavy Industries (YHCHI), in which the reader/viewer is forced to consume (a fact of which YHCHI are well aware and that they even self-reflectively build into their works) rather than work, such as "[theHouse]" by Mary Flanagan (which I discuss in chapter 4), that is built with Processing and so is open source, which makes it possible for reader/viewers to build on it or tinker with her poem and so create their own.[84] Perhaps, the difference is more stark when comparing an early work of digital poetry by Brian Kim Stefans, titled "The Dreamlife of Letters," that was built with Flash in about 2000 (shortly after Macromedia published a new version of Flash that included advanced actionscript) with a more recent work built with Processing, called "Letter Builder," that he released in the summer of 2009 (see Figure 30).[85] The latter piece, a DIY concrete-poetry builder, is something of a mirror for the code underneath—reading and writing the poem are complimentary processes based on a philosophy of making. (Also noteworthy, the aesthetic of Stefans's work is becoming less and less clean, more messy, and thus closer to the aesthetic of McCaffery's typescapes.)

What ties Processing to dirty concrete poems such as *Carnival*, as well as typewriter concrete poems by Nichol and Houédard, is a movement not only to democratize the creative process but also to combine this democratization with artworks that embody a self-reflexive sensibility that makes this democratization possible through techniques that draw attention to the art object as a created object—again, techniques that essentially turn the inside of the art object out through a philosophy of making. Alan Kay describes the computer skills one develops via programming as *distinct* from the skills one needs to develop print-based literacy: "In print writing, the tools you generate are rhetorical; they demonstrate and convince. In computer writing, the tools you generate are processes; they simulate and decide."[86] This last sentence is where I disagree with his vision of software literacy, for it is precisely the DIY philosophy as a means to achieving software literacy that underlies Processing, which in turn ties it to literary DIY typewriter dirty concrete such as *Carnival*—a work that is a kind of artist book that generates both rhetorical tools and processes by way of an activist media poetics.[87]

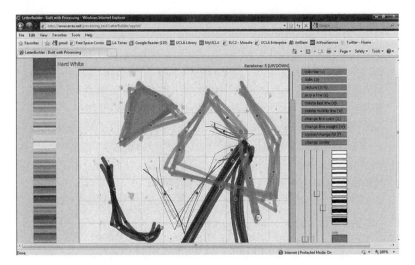

FIGURE 30. *Screenshot of Brian Kim Stefans's "Letter Builder" from 2009 (no longer available online).*

With its embrace of nonrepresentationality and illegibility via the typewriter, dirty concrete effectively communicates to us that in our current accelerated move into the supposedly clean digital age, the page, the typed letter, and the stamped word are media that we can (and should) be able to read *and* write just as much as we can computer software. Moreover, one of the ways in which Nichol, Houédard, and McCaffery communicate this is through hacking the page, the book, and the typewriter in order to renew them, to turn them from transparent carriers of meaning to objects meaningful in themselves.[88]

Dirty concrete poems are not an aberration in the history of twentieth-century poetry but rather representative of one of the mainstays of innovative writing: an active engagement in hacking both writing and writing media that treats both as process and product, the two unavoidably intertwined. Given this emphasis on making and doing, often through a kind of reverse engineering, it again seems clear that these works can be productively reread alongside the recent surge of digital DIY culture as a form of activist media poetry. McCaffery was prompted to write in the early 1970s, while he was at work on *Carnival*, "The typewriter oracled a neoclassical futurism that emerged in the mid twentieth century as poesie concrete. This is part of that oracle." But so too have these resolutely analog dirty concrete poems oracled our current cultural turn to the digital iteration of making as meaning—a turn that is strikingly exemplified by the underlying philosophy of different facets of the digital DIY movement, with the open-source programming language Processing as one particularly pertinent example.[89] Moreover, while as Michael Basinski writes, concrete poetry broadly "was so effective as an anti-academic, political tool, that it was exiled and abandoned and labeled by the machine as a trite form of outsider art" and "still waits for political poets to resurrect the form and make the charge anew," dirty concrete oracled digital media activist poems that *do* answer the charge anew.[90]

4

The Fascicle as Process and Product

> Windows, doors, airport gates, and other thresholds are
> those transparent devices that achieve more the less
> they do: for every moment of virtuosic immersion and
> connectivity, for every moment of volumetric delivery, of
> inopacity, the threshold becomes one notch more invisible,
> one notch more inoperable. As technology, the more a
> dioptric device erases the traces of its own functioning
> (in actually delivering the thing represented beyond),
> the more it succeeds in its functional mandate; yet this
> very achievement undercuts the ultimate goal: the more
> intuitive a device becomes, the more it risks falling out
> of media altogether, becoming as naturalized as air or as
> common as dirt.
>
> —Alexander Galloway, "The Unworkable Interface"

Against a Receding Present

Throughout this book I try to produce a friction from reading new
media interfaces with, into, and against old media interfaces—a
friction that not only troubles the distinction between new and
old but also follows in the steps of instances of (activist) media
poetics throughout the twentieth and twenty-first centuries
that similarly work against the grain of writing interfaces. This
chapter positions Emily Dickinson not only as a poet working
equally with and against the limits and the possibilities of pen/
pencil/paper as interface but also as one through which we can
productively read twenty-first-century digital literary texts.
My argument is that Emily Dickinson's nineteenth-century
fascicles—as much as mid-twentieth-century typewriters and
late twentieth- and twenty-first-century digital computers—
are now slowly but surely revealing themselves not just as media

but as media whose functioning depends on an interface that defines the nature of reading as much as writing.[1]

This chapter continues my attempt to read older and newer media against each other not to produce yet another neat, linear history of technological change but to instead *disrupt,* by swinging back and forth between analog and digital writing interfaces, contemporary computing's rapid acceleration of reducing creators as well as readers and writers into users and consumers whose access to the machine is limited to the surface gloss of a nearly invisible, supposedly intuitive interface. This back-and-forth bears the potential for disruption as it positions itself against the teleological narrative of technological improvement that is driving the disappearance of the digital computer interface under the guise of the user-friendly. As I explain in this section and touch on in chapter 1, the triumphant declaration that "the interface just disappears" is the logical end point of any version of the history of technology that charts its path into the future by tracking the movement from one invention to another while filled with nostalgia and marvel at the clunky, obvious materiality of, for example, pen and paper or the typewriter. Arguably, to invoke once more the ghost of Marshall McLuhan, these older analog media appear charming and clunky to us now because we are so enmeshed in our own media—media that work to veil their own workings. By revisiting older media, we can make our current media visible once again. Put slightly differently, my hope is that a media archaeological–inflected reading of the fascicle alongside and against the digital—one that firmly pushes against any inclination to say the fascicle is, for example, an early form of analog hypertext that anticipates digital hypertext—can *refamiliarize* the reading/writing interfaces we use every day so that we can look, once again, at our interfaces rather than through them.

Without reading the digital alongside and against the analog, the present slips from view, for the contemporary computing industry, which is accelerating its drive to achieve perfect invisibility, desires nothing more than to efface the interface altogether and so also efface our ability to read let alone write it. As Alex Galloway puts it in the chapter epigraph, the more user-friendly an interface is touted to be, the more invisible it is as it attempts to erase every trace of its own functioning.[2] An example of such an effacement occurred during one of the most well-known unveilings of a multitouch interface, at which creator Jeff Han proudly declared, "There's no instruction manual, the interface just sort of disappears."[3] Another example comes from the Natural User Interface Group, who define NUI as "an emerging concept in Human/Computer Interaction that refers to a interface that is effectively invisible, or becomes invisible to its user with successive learned interactions," and they use "natural" to mean "organic, unthinking, prompted by instinct."[4] But whose instinct is directing the shape of these interfaces? More to the point, why would we—whether we identify as a user or a creator, a reader, a writer—want our interactions with interfaces to be "unthinking," such that we have no sense of how the interface works on us, delimiting reading, writing, even thinking?

While this chapter clearly is historically inflected, using media archaeology as an underlying methodology offers a way out of the pitfalls of the term *history* and its orientation toward origins, along with its ideological weightedness toward linearity and uncovering, as opposed to the Foucaultian practice of history underpinning media archaeology that involves asynchronous cuts into the sedimentary layers of technological change. As such, this chapter is the last of four atemporal cuts concerned specifically with the interface. While critical theory has long disparaged master narratives and teleologies, there is still a deeply ingrained tendency to reproduce linear histories of media, as well as media writing such as digital literature (a field that continues to define itself, perhaps even legitimize itself, with origin stories about the history of computer-generated text from the 1950s and 1960s or the history of hypertext literature from the late 1980s and 1990s), and as such, media archaeology continues to be a much-needed critical intervention, since it reminds us that the study of media/literature no longer needs to involve un-covering a static series of firsts.[5] Instead, in the spirit of Michel Foucault's notion of the archive as a "system of statements (whether events or things)"—with an emphasis on "system" in-stead of the usual stress on "statements"—media history can be conceived as a shifting practice of uncovering the ways in which media themselves, in a very physical, concrete sense, engender and delimit what can be said, what can be thought.[6]

More, that the practice of media archaeology remains a shift-ing methodology and aesthetics is key, as any attempt to unify the study into a coherent or unified set of practices would not only inevitably reintroduce teleological narratives of media progress but also betray its Foucaultian roots in seeing archae-ology as the product of a never-ending process of describing the archive as past and present. As Wolfgang Ernst puts it, "Archaeology, as opposed to history, refers to what is actually there: *what has remained from the past in the present like archae-ological layers, operatively embedded in technologies*."[7] That said, media archaeology is less concerned with a Foucaultian empha-sis on power in social relationships and more concerned with the ways in which computer hardware—and for me, it is a con-cern that can be extended to writing media in general—exerts power over communication. In a sense, the reconfigured media archaeology approach I take here in relation to the fascicle is a reconfigured media archaeology applied both to a more distant past *and* to a constantly receding present that masquerades as the near future.

This chapter also marks one more attempt to move *interface* outside its conventional HCI, corporate-based definition—in which it is usually defined simply as the intermediary layer be-tween a user and a digital computer or computer program—and both apply it to writing media more broadly and reframe it as a transition point between the human reader/writer and what is/how it is written, a kind of threshold that, unlike windows and doors, does not simply lead from one space to another.[8] Interface-as-threshold is less a static, even neutral object or layer that allows a reader to interact with a machine and more an unde-fined point of access that is deeply in between human and ma-chine. More, interface-as-threshold gives us a sliding scale by which to assess the degree to which a given interface is more or less human or machine—as Galloway rightly points out, the de-gree to which an interface becomes more invisible is the degree to which it is seen as more user-friendly (and so more human), but at the cost of less access to the underlying flow of informa-tion or simply to the workings of the machine/medium.[9] Writ-ers such as Emily Dickinson who self-consciously tinker with both the reading and the writing interface are then performing necessary experiments with the "technological conditions of the sayable and thinkable."[10]

Moving the fields of HCI and literary studies closer together through a widening and reworking of the term *interface* does not signal a mere shift in terminology. Instead, hybridizing the two fields helps to move the study of literature—both bookbound and digital literature—into the post–Marshall McLuhan, en-abling us to go beyond repeatedly pointing out how the medium is the message and take up Katherine Hayles's well-received in-junction for "media-specific analysis" that gets at not just par-ticular media but particularities of the media, such as the inter-face, in the individual instantiations of literature.[11] This shift in focus means that once we read analog and digital writing media in terms of interface, we find that it no longer suffices to simply discuss a work and its page-level or screen-level effects as if either were created with a generic writing device, whether pen and paper, a book, or a computer. Furthermore, because digital interfaces in particular are so familiar to us now that they are de facto invisible, another underlying premise of this chapter is that it is necessary to look at older writing interfaces as a way to bring the digital back into view. Fascicles are, for example, obvious instances (because they often appear to be idiosyncratic) of the pen-and-paper interface, and so they too ineluctably frame what is and can be said just as much as the bound book, the typewriter, or the computer.

My Digital Dickinson

The relevance of a receding present for this section is that when reading Dickinson's fascicles through, alongside, and against the digital, first, the determining effects of the digital interface become clear, and then we see, by extension, not only that all reading and writing are determined by the interface but also that for analog *and* digital poets the interface needs to be open enough to remain both process and product—an intermediary layer between writing and reading that itself can be written and read. Here, I read digital literary texts into and out of the Dickinsonian fascicle, a writing interface that is both process and product from a past that's becoming ever more distant the more enmeshed in the digital we become and the more (fetishized) an object the book becomes. (If one requires proof that—so many years after Jerome McGann's *The Textual Condition*—the book has finally become an object, an object to be looked at rather than through, then one needs only to turn to the mainstream success of Jonathan Safran Foer and his 2010 *Tree of Codes,* an admittedly gorgeous die-cut book of erasure taken from Bruno Schulz's 1934 *The Street of Crocodiles.*)

Given an approach to media history that attempts to avoid simple teleologies, it is absurd to say that Emily Dickinson is a digital poet in the way we understand that term today—poetry that is *both* created using a digital computer and self-conscious and/or self-reflexive about its digital medium of creation and representation. It is equally absurd, because of its inbuilt determinism, to say that the variants in Emily Dickinson's work show that she was attempting to write digital/hypertext poems with the restrictions of pen and paper (i.e., if Dickinson could have written hypertext poems, she surely would have done so). But what if the approach were reframed slightly so that in addition to reading her fascicles and her variants as a kind of archive of mediated statements, we made the approach more literary? It is a literary approach less in the sense of inserting Dickinson into the present moment and arguing for her relevance to today's digital literary texts, such as those by Mary Flanagan, Aya Karpinska/Daniel C. Howe, and Judd Morrissey, and more in the sense of foregrounding the ways in which the digital now permeates our reading/writing habits, so that we read/write the present moment into Dickinson and argue for today's relevance to Dickinson.[12] That our reading should move *both* from the present into the past and from the past into the present— that we ought not to favor one approach over the other—is underscored by the fact that as I discuss throughout this book, writing technologies in general and digital writing media in particular not only cognitively change us as readers and writers but are constantly being remediated (in Jay Bolter's and Richard Grusin's still-relevant sense of the term). It is not just that we irremediably see the book through the lens of the digital but that the technology of the book finds its way into the digital— the book, reconfigured in our minds and in actual fact by the digital.

This back-and-forth between the book and the digital means that a media archaeology approach is not just an approach one *could* take to understand digital literary texts but if one approaches these works with any degree of historicity, an approach one *should* take. In fact, I argue that if we are to fully and accurately acknowledge the state of digital literature at the present moment, we will never successfully locate ourselves if we do not infuse our investigations into the contemporary with a sense of historical groundedness that at the same time is free from the teleologies I have discussed. Otherwise, it is all too easy to make claims about digital literature like those of Christopher Bantick, a journalist for the *Australian,* who declares, in passing, "[Jason Nelson's work] is engaging and what he has done with language is impressive. But while he is one of the new digital voices, he, along with lesser poets who rant and pant online, threaten the place of formal verse and structure."[13] Even though these claims are meant to be obviously inflammatory, they are still representative of many who are unsympathetic to digital writing of all kinds precisely because it *does* unmoor reading and writing practices that literature in English has held dear for centuries. As I iterate in this chapter and throughout this book, however, this is an unmooring with which writers working in a wide range of media have long been engaged.

It is undeniable that digital poetry, one genre among many in digital literature, is transforming the limits and the possibilities of poetry and poetics (so much so that unless the author specifies that their work is digital poetry, it's often unclear whether these works are poetry at all or are simply instances of digital literature). With respect to the work I discuss in this section, how do we as literary critics and scholars account for Mary Flanagan's "[the house]" or Aya Karpinska/Daniel C. Howe's "open.ended"? Both are engaged in representing the text as an emergent and explorable object—a three- and four-dimensional place in which to dwell that is simultaneously a material and a dematerialized place, one that is capable of visually reacting to the user's interactive struggle with the text. How do we account for works such as John Cayley's "translation" and "windsound," which exemplify his "ambient poetics"? In these works, the text unfolds over time with or without the participation of the reader, who can never quite grab hold of the text long enough to "read" it. How do we read Judd Morrissey's "The Jew's Daughter," which is similarly elusive as it invites readers to click on links embedded in the narrative text, links that do not lead anywhere so much as they unpredictably change some portion of the text before our eyes? Insofar as these (nearly but never quite tangible) texts are constantly changing, moving, generating, and emerging, not only do they seem to defy most conventions of literary texts and, even, of the most experimental poetry (for even a radical Language poem by, say, Bruce Andrews or Ron Silliman that aims to disrupt conventions of reader/writer/text is consistently the same text and can be returned to again and again), but an unbridgeable gulf seems to separate bookbound literature and these digital works. Where did they come from? Do they even belong to a literary lineage?[14]

To date, while there are abundant critical studies on digital film, digital archives, new media art, databases, hypertext fiction, artificial intelligence, artificial life, etc., the only book-length studies on digital poetry are Loss Pequeño Glazier's *Digital Poetics: The Making of E-Poetries*, published in 2001; Christopher Funkhouser's *Prehistoric Digital Poetry: An Archaeology of Forms*, a ten-year-long project published in June 2007; and his 2011 *New Directions in Digital Poetry*. All of these works position digital poetry in a lineage of avant-garde, modernist, and experimental writing traditions (ranging from Dada to Oulipo to Language Writing) in order to argue for the literariness, or the legitimacy, of digital poetry. Writing of the roots of digital poetry, Funkhouser declares:

> Digital poetry's foundations, mechanically and conceptually built in the decades *before* personal computers, were firmly established by the 1990s—*before* the WWW came into existence. This observation is significant . . . because the early history of this burgeoning genre is almost completely unknown, and the present state of digital poetry cannot be fully understood without a sense of its origins.[15]

He then goes on to write that "digital poets conceived of these works with the same poetic and theoretical practices used by artists who worked with nothing more than paper and ink" and that the "aesthetics of digital poetry are an extension of modernist techniques."[16] For Funkhouser, then, Williams and Pound are precursors to digital poetry in their use of juxtaposition, as are other "postatomic" writers who "use fragmentation to legitimize fragmentation and challenge the stability of language as a point of meaning."[17] While I agree that Williams and Pound may indeed be crucial digital poetry precursors, as long as we trace their influence according to broad formal and thematic techniques such as juxtaposition and fragmentation, then we are doomed to calling almost any poet going back thousands of years a digital poetry precursor. As such, on the one hand, at this early stage of defining the field of digital poetry or digital literature, any historicizing is much needed. On the other hand, we must be wary of too easily seeing literary precedents everywhere we look—to do so is to gloss over the defining effects of different writing media on the reading/writing experience. Digital poets may have indeed conceived their works "with the same poetic and theoretical practices used by artists who worked with nothing more than paper and ink," but equally or even more important, digital poets conceive their works on and for the fundamentally different medium of the computer/screen. It is a difference that makes a difference—although lest we consign digital texts as "other" in perpetuity, the difference between the digital and the bookbound is one that we should not be deceived into seeing as wholly unbridgeable, radical, untranslatable.

It also seems clear that if we can trace specific formal and thematic qualities of digital poetry back to modernism, then we most certainly can cross the divide separating the twentieth and the nineteenth centuries and trace these qualities back to writers such as Emily Dickinson.[18] In fact, if we take Dickinson as a test case, reading the digital into and out of Dickinson may enrich our understanding of her work. We can self-consciously exploit the terminology and the theoretical framing of the present moment, which—given the ubiquity of born-digital terms such as *interface, network,* and *link* or, even, of now commonly understood terms such as *bookmark* and *archive,* which were previously used only by the bookish or the literary scholar—is steeped in the digital and which often without our knowing saturates our language and our habits of thought.

Surely, a self-consciousness about and strategic exploitation of the structures built into our everyday digital computing will also reinvigorate the terminology and the theoretical framework we use to understand, for example, Dickinson's variants.[19] Thus, Sharon Cameron's highly influential descriptions of the variants, which in *Choosing Not Choosing* are infused with the language of critical theory so popular in the 1990s (e.g., variants are described as forms of identity, as heteroglossic, etc.), can be seen anew and even augmented with our current sense of variants as multidimensional, spatiotemporal linkages.[20] What I am proposing is certainly neither new nor groundbreaking. For example, Martha Nell Smith—the executive editor of the *Dickinson Electronic Archives*—directly stated in 2002, "New media challenge us to consider what can be gained by amplifying our critical commentary into more media and how our critical-theoretical tools can be shaped to exploit multimedia most effectively."[21] Given the relative paucity of critical writing on Dickinson in relation to the digital, however, this chapter is my attempt at taking up the challenge of using digital media to read earlier bookbound poets into and out of digital poetry/ literature.

This mode of reading that explicitly and, again, self-consciously uses the present to read the past makes possible, for example, the retrospective observation that our appreciation of Dickinson is a direct result of the 1981 publication of R. W. Franklin's *Manuscript Books of Emily Dickinson*.[22] We have only recently come to see the Dickinson who pins together scraps and creates collages of sorts from fragments written at angles to each other, the Dickinson of variants and multiple versions, the Dickinson who is acutely aware of pen and paper as a technology, as a writing media. It is worth noting that 1981 was also the year that IBM released their first mass-market, affordable personal computer (PC), the IBM 5150 (as I discuss in chapter 2, 1981 was also the year Xerox released the 8010 Star Information System, the first commercially available computer to use a GUI). Within only a year or so of the release of the PC, as Funkhouser's informative timeline indicates, the creation of poems mediated and modulated by a computer and/or computer screen had doubled. Looking back, it cannot be insignificant that poems such as *First Screening* by the Canadian poet bpNichol introduced us to new forms of reading/writing that were simultaneously kinetic, multidimensional, spatial, and temporal—forms of reading that may also inform our reading of Franklin's *Manuscript Books*.

Furthermore, doing such archeological exploration would enrich our understanding of digital poetry/literature itself such that we could not so easily claim these works were examples merely of, as Marjorie Perloff puts it, "techniques whereby letters and words can move around the screen, break up, and reassemble, or whereby the reader/viewer can decide by a mere click to reformat the electronic text or which part of it to access."[23] While these digital works do indeed "become merely tedious unless the poetry in question is, in Ezra Pound's words, 'charged with meaning,'" we need to learn to become more perceptive readers of digital poetry/literature so that we do not simply frame every piece with bookbound assumptions driven by the equally bookbound practice of close reading.[24] In other words, while we know what "charged with meaning" looks like in a poem by Dickinson, Williams, or Pound, it is not a given what "charged with meaning" looks like in the digital. Further, a fine-tuned digital poetry literacy would also prevent us from allowing ourselves to be deceived into thinking that digital literature was an abrupt break from the bookbound. Dickinson could very well be a, not the, mother of them all.

Antidote to the Interface-Free

Even the meaning of the expression "the mother of them all" has been transformed by digital computing. Kabbalah scholars would recognize the phrase as coming from the *Zohar,* or the phrase might recall Saddam Hussein's 1991 reference to the Persian Gulf War as being "the mother of all battles." Those working in the IT industry use the phrase, however, to refer to Douglas Engelbart's groundbreaking demo from 1968, in which he presented, for the first time, his invention of the keyboard-screen-mouse (KSM) interface, as well as introduced teleconferencing, videoconferencing, e-mail, and an early form of hypertext. Now, the KSM is so seamlessly a part of our everyday work and leisure—mediating and defining most acts of writing, reading, and thinking—that we no longer notice it as an interface at all. Steven Johnson writes in his 1997 *Interface Culture* that we need to start developing criteria by which to judge our interfaces: "If the interface medium is indeed headed toward the breadth and complexity of genuine art, then we are going to need a new language to describe it, a new critical vocabulary."[25] I doubt he could have envisioned, however, that ten years later we as a culture would not only remain largely oblivious to the way interfaces work on us, rather than most of us working on them, but be poised to begin an era not of the interface as art but of the interface-free. In fact, only a brief survey of interface-design guides from the past decade clearly indicates the ways in which mainstream interface design, or even the education of interface designers, is fully entrenched in moving as far as possible from Johnson's vision of the interface as art and toward a naturalized or invisible interface. Take, for instance, Joel Spolsky's *User Interface Design for Programmers,* which states at the end of chapter 8:

1. Design for people who can't read.
2. Design for people who can't use a mouse.
3. Design for people who have such bad memories they would forget their own *name* if it weren't embossed on their American Express.[26]

Or take a more contemporary example by Janet Murray, who unlike Spolsky identifies as both a scholar and a humanist. In her 2011 *Inventing the Medium: Principles of Interaction Design as a Cultural Practice,* she takes up McLuhan's belief that media are extensions of ourselves but neglects the crucial point that *media are also inherently ideological,* which he makes via his criticism of General Sarnoff's assertion that tools are neither good nor bad—it's what one does with a tool that matters. While she rightly criticizes the popular reliance on the term *intuitive* to describe a successful interface—"'intuitive' is by far the most abused word in digital design and it is one that should perhaps be banned for a decade or so until it can once more be employed

meaningfully"—in the end Murray counsels designers to instead aim for transparency: "In order to make truly intuitive
interfaces, designers must be hyperaware of the conventions
by which we make sense of the world" so that they can instead
design an interface that "does not call attention to itself, but . . .
[lets] us direct our attention to the task."[27] This particular notion of transparency is positioned admirably against thinking
of computers as "black boxes or limited function appliances,"
for as Murray writes, *"The more visible we can make the operations of the machine, the more control we can give to the expressive user, and the more we can foster the development of expressive
technique."*[28] Designers will never produce transparent tools,
however, as long as they see interfaces as "extensions of the
hand." Trapped in the limited confines of an anthropocentric
model of technology that cannot account for the ways in which
computers are at least in part resolutely nonhuman, this kind
of drive to transparency can only ever more deeply conceal the
workings of the interface, workings that are neither neutral
(and so cannot simply be revealed through a model of the interface as window) nor entirely humanlike.[29] Instead, the model
of interface-as-threshold reminds us that there is no getting
beyond or outside the interface and of the ways in which it is
equally human and machine.

If work by media studies theorists does not yet fully reso-
nate with us, then experimental writing as a mode of media
poetics shows us a way out of this anthropocentric trap of
black-boxing our machines even as we endeavor to make their
workings transparent. More specifically, one of the most important lessons to take away from the manuscripts and the
editing of Emily Dickinson is that there is no such thing as
the interface-free—that it is absolutely necessary we both acknowledge that all writing comes to us through an interface
and identify the precise ways in which the interface, whether
pencil/pen/paper or KSM, inescapably defines or frames such
writing. While I agree with Cristanne Miller's assertion that
"Dickinson's poetry can only accurately be read when freed

from the constraints of conventional print typography and conventional conceptions of her poems," I do not believe it is possible to have access to a pure reading of Dickinson's poems, one that is unmediated either by twentieth- or twenty-first-century reading/writing interfaces or by our own thinking habits, which are similarly enmeshed in reading/writing interfaces.[30] Henry Petroski, author of *The Pencil: A History of Design and Circumstances,* points out that understanding the development of writing interfaces such as the pencil (or the pen) "helps us to understand also the development of even so sophisticated a product of modern high technology as the electronic computer."[31] The cost of ignoring what Dickinson teaches us about writing interfaces is abundantly illustrated in Susan Howe's and others' work on the limitations of relying solely on Thomas H. Johnson's *The Poems of Emily Dickinson,* which utterly ignores Dickinson's writing interfaces and instead problematically reframes her work with the interface of the printed book and the conventions of typography. As Franklin more mildly puts it in the introduction to his facsimile edition of the fascicles, "The variorum edition . . . edited by Thomas H. Johnson, translated the mechanics of the poems into conventional type and, in presenting them chronologically, obscured the fascicle structure. Such an edition, though essential, does not serve the same purposes as a facsimile of the fascicles."[32]

To return for a moment to the recent turn to the interface-free and the pressing need to read writing interfaces, in February 2006, NYU research scientist Jeff Han unveiled to attendees at the O'Reilly Emerging Technology Conference the first affordable version of what he called an interface-free, touch-driven computer screen. Shaped like a thirty-six-inch-wide drafting table, the screen allowed the user to perform almost any computer-driven operation through "multi-touch sensing" that was, as Han described it, *"completely intuitive.* . . . There's no instruction manual, the interface just *sort of* disappears."[33] The response was and continues to be unqualifiedly enthusiastic. As one audience member put it, "To see it is to be blown away by its simplicity and elegance." I attach significance not only to his phrase "completely intuitive" (which prompts me to ask, just whose intuition is driving this interface-free interface?) but also to his qualification "sort of" (as in it "sort of disappears"). No doubt, the interface-free system Han proposed is elegant, beautiful, compelling—like walking into a gleaming white-and-chrome Mac store. But after the initial pangs of longing pass for this newest of the new, I am left wondering why we continue to long for this sort of false transparency? Why do we lure ourselves into believing that these interfaces offer us the ability to somehow transcend the interface itself and not understand that they instead offer us an increasingly difficult to pin down, perhaps even insidious form of control on our creative expression? As Lev Manovich reminds (or warns) us, "The interface shapes how the computer user conceives of the computer itself. It also determines how users think of any media object accessed via a computer. . . . In short, far from being a transparent window into the data inside a computer, the interface brings with it strong messages of its own."[34]

As critics such as Susan Howe, Marta Werner, Jerome McGann, and Martha Nell Smith make clear, coming long before Marshall McLuhan's famous dictum that "the medium is the message" and long before the emergence of the term *interface* in the 1960s, which at that time referred only to the interaction between two systems, Emily Dickinson was exemplary in her keen awareness of the limits and the possibilities of the writing interfaces of her time. Indeed, "the very commonness of the pencil" (or the pen) is what "renders it all but invisible and seemingly valueless" such that it "becomes a part of society and culture so naturally that a special effort is required to notice it."[35] By contrast, Dickinson understood pen/pencil/paper as an interface—she was acutely aware of the limits and the possibilities of the triad such that "shapes and letters pun on and play with each other. Messages are delivered by marks."[36] Nowhere is this understanding of the writing interface more evident than in her pinned poems, especially those she created after she turned away from the book-inspired form of the fascicle in 1864.

In a note at the end of the second volume of *The Manuscript Books of Emily Dickinson,* Franklin claims that Dickinson's practice of pinning was one of several methods she used when she needed to add extra lines. He writes, "Early in 1862 she pinned slips to accommodate overflow when she reached the end of a sheet, but she came to favor another way: a separate sheet carrying only the additional lines. . . . When ED ceased binding fascicle sheets, about 1864, she reverted to pinning slips to sheets to maintain the proper association."[37] But Dickinson had to have been doing much more than pinning an extra sheet to establish a relationship between the content of the two pieces of paper. That is, the manuscript version of a poem such as "We met as Sparks" can be read as an instance of Dickinson's desire to draw attention to the mediating effects of pen and paper, and therefore poems such as this one are also attempts to both denaturalize the writing media and disrupt our tendency to see *through* the writing surface (or simply to not see the writing surface at all).[38] First dated 1864 by Franklin and then later changed to 1865, this poem appears on the verso of Set 5, designated A 92–14 (see Figure 31). Two additional (or alternate) lines (or perhaps three lines, depending on your position on Dickinson's intentions regarding her line breaks) are pinned to the bottom of the poem so that the final lines of the poem proper are covered. Not surprisingly, however, the version of "We met as Sparks" in Franklin's 1998 variorum edition, which is translated into the typographic regularity of the page, is stripped not only of its riveting physicality but also of this particular discourse on writing media that is self-consciously expressed through writing media. It is as if it is not the same poem at all.

First, the manuscript version of the poem shows a writer who has a precise understanding of the dimensions of the page—in fact, given that she writes a consistent distance from both the left and the right edges of the page, it appears not only as though she has a painter's sense of the shape and the size of her letters and words, the size and the shape of the page as a canvas, but as though her line breaks are entirely intentional. No typeface or typographical spacing can adequately translate the handwritten word—it certainly cannot express the particular shape of the letter *S*, for example, that is echoed across the page to visually and aurally associate "Sparks" with "Sent," "scattered" (note the lowercase *s*, which is a sort of literal representation of scattering), "Subsisting," and finally "Spark." Nonetheless, in the following poem, I transcribe the poem with the line breaks as they appear in the manuscript version. The version without the pinning is on the left, and the version with the pinning, which covers, rewrites, or writes over the final three lines of the poem, is on the right:

We met as Sparks—	We met as Sparks—
Diverging Flints	Diverging Flints
Sent various—scattered	Sent various—scattered
ways—	ways—
We parted as the	We parted as the
Central Flint	Central Flint
Were cloven with an	Were cloven with an
Adze—	Adze—
Subsisting on the Light	Subsisting on the Light
We bore	We bore
Before We felt the	Before We felt the
	Dark—
Dark—	A Flint unto this Day—
We knew by change	perhaps—
between itself	But for that single Spark.
And that ethereal	
Spark.	

Contrary to Walter Benn Michaels' declaration in *The Shape of the Signifier* that once we treat everything in a Dickinson poem as meaningful, then nothing is meaningful, here we are presented with a poem where everything indeed seems to contribute to the poem as an extremely complex, multifaceted object or a William Carlos Williams–esque poem-as-machine.[39] Right at the beginning of the poem, both the line break, which creates a small space of blankness, at the end of "We met as Sparks—" and the dash serve to dramatize the sudden movement of an ignited fleck into the air. Further, the separation of "Diverging Flints" from the first line not only similarly dramatizes divergence but also introduces another connotation of "spark": unlike the definition of a spark as that which singularly erupts and disappears into the air, the line break seems to suggest the definition of spark as the "luminous disruptive electrical discharge of very short duration between *two* conductors separated by a gas (as air)."[40] In fact, meeting as sparks while "Diverging as Flints" (and then "cloven with an / Adze") expresses not only the tension Dickinson explores throughout the poem of the way in which any coming together involves simultaneously a merging, a certain loss of singularity, and an inevitable sense of separateness that can never be overcome but also that the "we" of the poem (and note that there is only a "we" and never an "I") is both the catalyst (i.e., the flint) and the thing catalyzed (i.e., the spark).

Throughout the poem Dickinson continues using such techniques of enjambment and of merging the literal and the metaphorical with the physical dimensions of words. For example, it does not seem coincidental that the line break separating "Sent various—scattered" from "ways—" enacts a scattering, as the eye must move from one side of the page down to the other. It does not seem coincidental that, on the one hand, the version of the poem on the left begins with "We met as Sparks" and ends with the singular "Spark" on a line by itself and that, on the other hand, the version on the right, with the pinning, replaces or *changes* the lines underneath (which read "We knew by change / between itself / And that ethereal / Spark") with "A Flint unto this Day— / perhaps— / But for that single Spark." Note that a sense of uncertainty or of thinking poised between two conflicting positions is expressed in the word "perhaps,"

which is placed on a line by itself, as well as by the reference to
a singular spark, which in this case does not appear on a line by
itself. The poem is the version on the left at the same time it is
the version on the right. The poem is the version on the left, *or*
it is the version on the right. It is the former at the same time
it is the latter. It is about (the tension inherent to) singleness
and doubleness as much as it physically manifests itself as both
single and double.

There is also a temporariness to the pinning in the same way
that clothing is pinned either as a form of temporary stitching
or as a way to mark where the fabric may later be sewn. The slip
of paper has been pinned, not sewn, to the sheet of paper, and
so it is simultaneously bound and unbound. As Marta Werner
writes of another, later pinned poem, designated A 821, "The
pin complicates the play among past, present, and future. . . .
For here, the expectations of closure or *parousia*. . . may be end-
lessly postponed, or reversed, with the drop of a pin." She further points out the distinctiveness of pinning, for "unlike binding, which is premeditated, permanent, and serial, pinning is
instantaneous, temporary, random."[41] As such, the pinning in
"We met as Sparks" is so much more than an instance of Dickinson writing "the alternative on a slip of paper" as a way to "complete the poem"—the pinning makes impossible any reading of
the poem as complete.[42]

Also, Dickinson's handwriting in this poem, her use of the
space of the page, is a formal and thematic element of the poem
itself and so is untranslatable into any other medium. The
poem is absolutely and self-consciously of its writing medium,
and thus, to read Dickinson into the present moment, her work
refuses to let us be seduced by the idea of the interface-free.
It is a clear invocation for us to stay away from claims such as
those made by Jaishree Odin, who in writing about the digital poet Stephanie Strickland, declares, "Unlike the print
medium where content is the same as the interface, the database produced by the writer for the digital medium needs an
interface to make it accessible to the user. For the first time

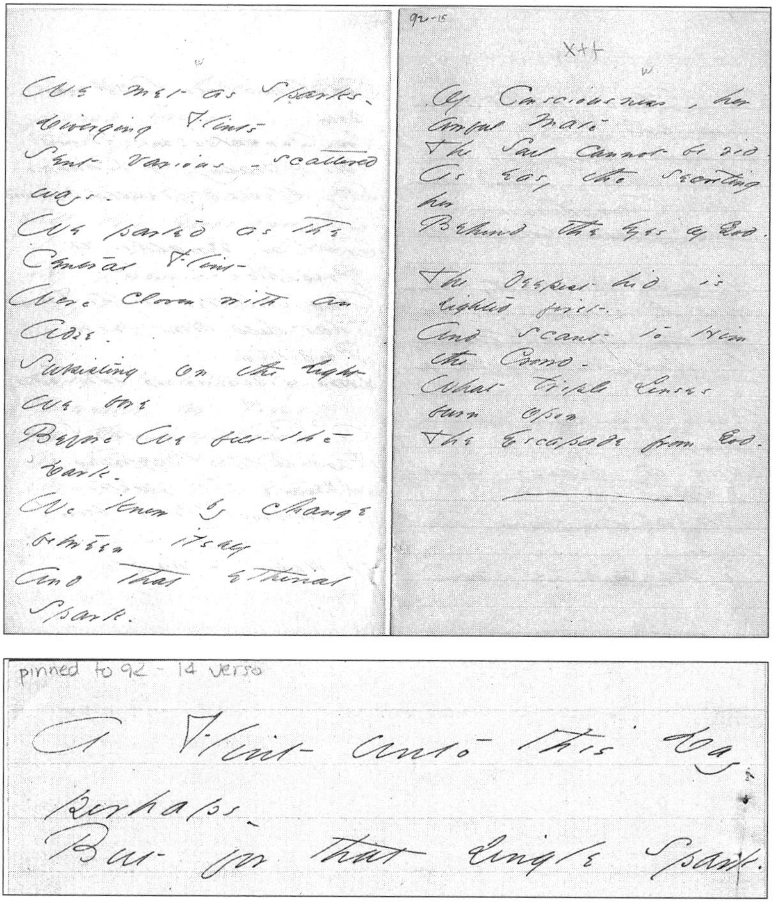
We met as Sparks –
Diverging Flints
Sent various – scattered
ways –
We parted as the
Central Flint
Were cloven with an
Adze –
Subsisting on the Light
We bore
Before We felt the
Dark –
We knew by change
between itself
And that Etherial
Spark.

92-15
X+F
A Consciousness, her
awful Mate
The Soul cannot be rid –
As easy the secreting
her
Behind the Eyes of God.
The Deepest hid is
sighted first –
And scant to Him
the Crowd –
What triple Lenses
burn upon
The Escapade from God.

pinned to 92 - 14 verso
A Flint unto this Day,
perhaps,
But for that single Spark.

FIGURE 31. *Facsimile reproduction of Emily Dickinson's pinned poem A 92–14a, "We met as Sparks—Diverging Flints," and the pinning on the verso of the poem. Courtesy of the Amherst College Archives and Special Collections.*

we have a distinction between the content of the work and the interface to access it."[43] Nowhere is the mistakenness of Odin's claim that "content is the same as the interface" in the print medium more evident than when we try to translate a Dickinson poem either into the typographical page, as I do here, or, as have Thomas Johnson and F. W. Franklin, onto the computer screen. There's no doubt that the version of "We met as Sparks" in Franklin's variorum edition is a neater, tidier poem. With both "We met as Sparks—Diverging Flints" and "Sent various—scattered way—" as one line rather than two, all of the lines are consistently the same length. Also, breaking the second line at "ways" rather than "scattered" adds a more orderly dimension to the poem, as *scattering* refers to random or chaotic movement, whereas *ways* refers to predetermined directions. Further, coming at the ends of lines one and three, the repetition of "Flint" is more obvious, as is the end rhyme between "Dark" and "Spark" in lines six and eight. Perhaps far more important is that the sense of the poem—the content of the poem as much as its physical structure, the poem as a material artifact that is simultaneously single and double—is occluded so that "We met as Sparks" is flat and hierarchical: a primary text is supported by a secondary, less important set of alternative lines and an even less important (indicated by the small font size) list of the line breaks as they appear in the manuscript.[44] Of course, such a hierarchy of primary text and alternate lines and line breaks is entirely excluded from the far more affordable and therefore far more commonly used reading edition of Franklin's *The Poems of Emily Dickinson*.

Thinkertoys

To read/write the present moment into Dickinson and argue for today's relevance to Dickinson, we should see her pinnings as well as her variants not so much as bookbound examples of chunk-style hypertext (links that allow the reader/user to move from one page to another, the kind of hypertext that is almost entirely responsible for the current structure of the Internet) but as *thinkertoys*. Before I explain the meaning of this term, I would point out that while calling Dickinson's pinnings or variants forms of hypertext does draw attention to the physical separateness of and connection to each word or chunk of text, the hypertext we usually use on the Web is directional and even linear. Links on the Web move us through the text or a series of texts in ways predetermined by the writer/programmer and, therefore, quite unlike how pinning functions in "We met as Sparks." In it, pinning makes the poem both two-texts-as-one-text and two separate texts. This simultaneously single and double nature of her work cannot be replicated online, but a Web-based translation of "We met as Sparks" could be created by linking together scanned images of the sheet and the pinning, layering the one over the other. Given the reading experience that is fostered by the KSM interface, which is entirely different from that of the book, such a translation would have to be approached as a thinkertoy.

The term *thinkertoy* was coined by Theodor Nelson in his 1987 *Computer Lib / Dream Machines*. He writes:

> Our greatest problems involve thinking and the visualization of complexity. By "Thinkertoy" I mean, first of all, a system to help people think. . . . But a Thinkertoy is something quite specific: I define it as a computer display system that *helps you envision complex alternatives.* . . . We will stress here some of the uses of these systems for handling *text*. . . partly because the complexity and subtlety of this problem has got to be better understood: the written word is nothing less than the tracks left by the mind, and so we are really talking about screen systems for handling ideas, in all their complexity.[45]

Therefore, instead of placing the emphasis on the production of new editions, versions, or translations of Dickinson's manuscript poems, we could emphasize the ways in which either a given reading/writing interface or a set of conceptual terms belonging to an era of a reading/writing interface allowed us to think more expansively about the work at hand—to better map the multilayered intricacies of a given poem. Hypertext or any other digital mode of representation would then become less a "radically new information technology" that wholly disrupted our notions of reader/writer/text and more another technology by which to reposition ourselves in relation to the reader/writer/text.[46] To read hypertext or any other digital writing media into and out of Dickinson means that digital writing such as Mary Flanagan's "[the house]," Aya Karpinska/Daniel C. Howe's "open.ended," and Judd Morrissey's "The Jew's Daughter" are no longer only an instance of a foreign, textual object of fascination but are simultaneously readable and unreadable, intimate and other, variable and static. (Such digital writing is also a textual instantiation of an ongoing poetic exploration both of the specific limits and possibilities of the space/time of writing and of language as an elusive yet multidimensional dwelling space. These digital works, not unlike "We met as Sparks," are ineluctably both this and that.) To read and think through Dickinson's work is, then, to be prepared for these stubborn, uncomfortable works that are simultaneously single and double material artifacts stubbornly mediated by reading/writing interfaces.

First, Mary Flanagan's "[the house]" is a digital poem environment that consists of strings of transparent, three-dimensional, occasionally intersecting, yet shifting boxes accompanied by paired lines, which in turn are recombined as the piece progresses, that we may watch as they move across the screen, grow larger or smaller, and rotate so that we read them in reverse—as if we could walk to the back of our language (see Figure 32).[47] Or should we want to actively determine the shape and the direction of the text/boxes, we may try to interact with the text/boxes through the mouse. Flanagan writes that "as in much of electronic literature, the experience of the work as an intimate, interactive, screen-based piece is essential to understanding and appreciating it," and should we choose to interact with this text environment, the experience is primarily one of struggle or difficulty, as there is no way to gain control over the text—no way to determine the direction in which the piece shifts. Pulling right on the mouse does not guarantee that the text will also shift right or rotate clockwise, and moving

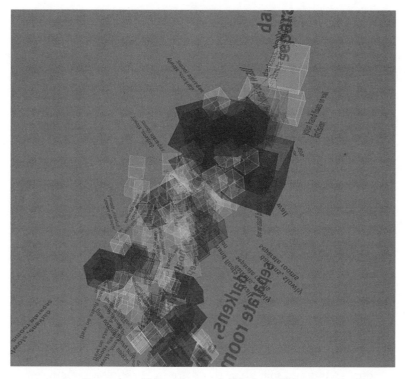

FIGURE 32. *Screenshot of Mary Flanagan's "[the house]" from 2006.*

the mouse up similarly does not necessarily allow us to venture deep inside the boxes or the text—we may have just flipped the boxes/text or moved to a bird's-eye view of this strange computer-text-organism. Thus, despite my interactions with the text, despite the fact that I can "read" most of the lines, in its difficulty "[the house]" is at least in part *about* the mediating effects of an interface that despite Flanagan's claim, offers at the same time it declines intimacy. Certainly, this work embodies the complexity, the possibility to explore complexity, that defines Nelson's thinkertoy.

But what of close reading in the digital realm? My reading of Dickinson's "We met as Sparks" involves a fine attention to the aural and the visual structures of the poem. Even though

the poem has an inbuilt aspect of instability because of the pinning, each time I read it, however, don't I return to the same
object, the same text again and again? I would answer that the
multiplicity of "[the house]" reminds us that there are at least
six different versions of "We met as Sparks." The first version
would include the recto *and* the verso; the second version would
include the recto, the verso, and the pinning as an alternative
ending; the third would include the recto, the verso, and the
pinning as an additional ending, and so on. Should we decide to
take into account the individual reading experiences we bring
to the poem and depending on whether we rely on a facsimile
version, there are no doubt many more versions than just six.
Thus, "We met as Sparks" denies closure or definitive reading in
the same way that "[the house]" does. It is just that the conventions of the book, in which we have been so enmeshed, lull us
into believing that a paper-bound or bookbound text is stable,
perhaps even knowable.

Further, reading "We met as Sparks" alongside "[the house]"
brings to light the ways in which the interface of each poem
bears with it a different set of standards for reading. Whereas
some sound and visual patterns are in Dickinson's poem, for
example, Flanagan's work has no aural element, and the visual
structure is not down or across a page or a sheet of paper but is
a rotation in and around a virtual three-dimensional space. Despite the variability of Dickinson's poem, I am still able to quote
from it, whereas with Flanagan's I *could* quote some (certainly
not all) of the different and recombined lines (for example, "giving emptiness / letters have their sharpness" or "the study almost finished / mouth to tell me"), but what would be the point,
especially when we cannot read the whole text or even know
where the text begins and ends? Is this text in fact many, many
texts that ought to be differentiated from each other in terms
of time rather than space? Thus, rather than asking ourselves
whether the poem on the one side of the page or the sheet of
paper is separate from the poem on the opposite side, the question changes to whether the text we see at five seconds into the

poem viewing is a separate poem from the text we see after two minutes of viewing. If we then interact with the text, as Flanagan encourages us to do, we have before us a nearly limitless number of different texts and different reading experiences.

But simply because we cannot read Flanagan's poem in the same way we read Dickinson's does not mean that the former is not a poem. Rather, both demand we find that point in the text where our reading practices fail us. It is at that point of failure that we may begin attending to the particularities of *the event* of each poem—the original event of the physical writing of the poem that took place through a particular interface, the event of our readings of the poem that take place through yet other particular interfaces—and begin taking an account of what is gained and lost through each mediation. Thus, I would like to end this chapter with various attempts at readings or accountings of gains and losses in works by Karpinska/Howe and Morrissey.

Although Aya Karpinska and Daniel C. Howe's "open.ended" offers itself as a poem environment in which to think through the possibilities offered by three- or four-dimensional writing, I would argue that unlike Flanagan's "[the house]," there is no resistance in it that might indicate an interface self-consciousness.[48] The authors describe their work as follows:

> With real-time 3D rendering & dynamic text generation, *open.ended* attempts to refigure the poetic experience through spatialization & interaction. As visitors manipulate a joystick to control interlocking geometric surfaces, stanzas, lines, & words move slowly in & out of focus, while dynamically updating text maintains semantic coherence. Order is deliberately ambiguous & multiple readings encouraged as meaning is actively & spatially constructed in collaborative fashion & new potentials for juxtaposition, association & interpretation are revealed.[49]

Here, reading merges with both viewing and navigation, as we must move either of the two sliders at the bottom of the screen in order to rotate and read the text on the walls of the outer or inner cubes at our own pace (see Figure 33). Or we could decline to read altogether by rotating the outer cube with the mouse so that we were peering inside the transparent walls of the rotating cubes. Or as the poem cube rotates without our interacting with it, we could read/view the poem according to its predetermined pace of reading. Still, though, the text behaves exactly as we expect it would—it consistently rotates on its own in the same direction, and our interaction with it obeys the usual rules of the KSM interface (e.g., moving the mouse right turns the poem cube right)—which results in a poem whose form is more an add-on or a technical feature than an intricate extension or reflection of its content.

Furthermore, what Karpinska and Howe do not mention is that the piece includes a looped audio recording of a male and a female voice reading some, but not all, of the text (sometimes consisting of recombinations of lines from different walls) that appears on the cubes. The effect is one of moving from a poem whose meaning may indeed be "deliberately ambiguous" and multiple, as the authors put it, to one whose meaning seems quite clearly to be about a physical coming together of the man and the woman. Their spoken words (phrases such as "GET / EMOTIONALLY / UNDRESSED," "EYES CLOSED / I AM / ANYWHERE," and "AN INSATIABLE / NEED TO / REPEAT") first separate and then gradually, increasingly overlap and intertwine as the recording progresses, ending on the phrase, "WE UNFOLD THIS FANTASY AND SURRENDER." Although the authors' attempt to spatialize and materialize the reading/ writing experience (by way of two intersecting three- or four-dimensional writing surfaces) gestures toward Dickinson's variants and her pinnings, bringing Dickinson's writing to bear on "open.ended" reveals potential for much greater conceptual and linguistic sophistication.

In chapter 1, I mention Judd Morrissey's "The Jew's Daughter" in relation to a poetics of failure in digital literature and its accompanying limits of interpretation, and here I would like to

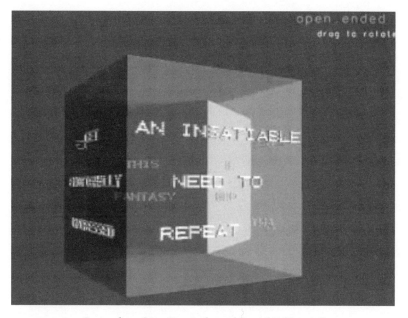

FIGURE 33. *Screenshot of Aya Karpinska and Daniel C. Howe's "open.
ended" from 2004.*

broaden my reading to account for the work's overall complex
relationship to the bookbound page—the way in which it reads
and reworks both the bookbound page through the digital
and the digital through the bookbound page, a self-conscious
doubleness that reads its own writing interface in much the
same way as do Dickinson's pinned poems.[50]

Morrissey describes "The Jew's Daughter" as "an interactive,
non-linear, multivalent narrative, a storyspace that is unstable
but nonetheless remains organically intact, progressively weav-
ing itself together by way of subtle transformations on a single
virtual page."[51] "The Jew's Daughter" consists of roughly 608
pages of recombinant chunks of texts, and indeed, "pages" is
more than a skeuomorph, as each screen of text—a white rect-
angle with mostly black text—intentionally emulates a page in
a book. It *is* possible to read the text on each page/screen from
beginning to end, left to right, as one would a page in a book,

but this is reading, in terms set by the book, in the most limited sense, as the difference is in the way this reading prohibits any kind of physical interaction with the text. We may always finger a bookbound page, hold the page as we anticipate turning it, fold over the corner of or underline passages from a particularly provocative page. But as each page includes one word, letter, or character that appears in blue, much like a standard hypertext link, the text on a given page can be read from beginning to end only if you refuse to touch or interact with the text in any way. Further, whereas the hyperlinks we are accustomed to using on the Internet take us to a new page, one whose subject matter is clearly related (at least in the mind of the coder) to the original page, these links neither are clickable nor take us to a new page, leaving the old page intact. They are, instead, *temporal* linkages. Running your mouse over the blue word activates the Flash programming and results in the disappearance/replacement of random chunk(s) of text. From one page to the next, the reader can never predict how, where, or why the text has changed. Thus, as Katherine Hayles points out, reading becomes an act of memorization, as you need to be able to visualize and/or memorize the content of the first page in order to know what has changed or in order to read the text in the manner to which we are accustomed. As she puts it:

> When the player mouses over the blue letters, some part of the text, moving faster than the eye can catch, is replaced. Reading thus necessarily proceeds as rereading and remembering, for to locate the new portion of the page, the reader must recall the screen's previous instantiation while scanning to identify the new portion, the injection of which creates a new context for the remaining text.[52]

In other words, Morrissey has created a temporally based palimpsest in that chunks of texts are layered on top of each other (but only in the reader's mind) as the text is unfolded over time. From one page to the next, some text stays the same and so in a sense remains legible, whereas other chunks of text are replaced, reworking the meaning of both the text that stays behind in the reader's memory and the text that is still visible. It is conceivable that "The Jew's Daughter" is not 608 individual combinations of text chunks but rather a piece of conceptual writing that challenges the reader to mentally assemble all 608 pages into a single text whose meaning does not reside in any one page. For example, on the first page one can see that the vague references to the activities of "she," "I," and "you" result in a determinedly indeterminate text that is not particularly about anything (see Figure 34a). The text could be about a "she," "I," and "you," but these pronouns also could be read as stand-ins for a commentary on the text itself—for wouldn't the reader wonder, will she (or it) disappear? Likewise, the following could be read as confirmation of my reading of "The Jew's Daughter" as a single palimpsestic text that can be read or understood only cumulatively, over time: "To hand to you the consecrated sum of your gifts, the secret you imparted persistently and without knowledge, these expressions of your will that lured and, in a cumulative fashion became a message."

Given the way the text consistently comments on the book in order to comment on the digital in order to comment on the book, it is perhaps fitting that in the context of the content of the text, which may be read as a comment on itself, the "you" could be both the reader and the writer. That is, if the text is indeed indeterminate, then the writer is handing to the reader the gift of their own reading of the text, or if the text is only limitedly indeterminate, then the passage can be read as saying that the reader's reading of the text makes legible the writer's secrets, expressions of their will that are legible only through the reader's cumulative work. We can apply the same interpretative technique to the sentence, "You asked could I build you from a pile of anonymous limbs and parts," and ask ourselves whether the writer (or the "I") is writing about the act of writing, of compiling a coherent text from a "heap of language" (to invoke Robert Smithson), or whether, with some pronoun slippage, it is the reader who must build the text from the writer's 608 pages of scraps of text. The way in which "The Jew's Daughter" tends to double itself, commenting on the reader/writer/text triad from as many perspectives as possible, is only reinforced by subsequent pages. For example, after running your mouse over "criminal" on the first page (is this word in particular meaningful, as criminals turn themselves against the law or even turn the law against itself?) and then reading the second page in relation to the first, the reader discovers that the sentence beginning "To hand to you the consecrated sum of your gifts" has been replaced with the following: "June through clouds like sculpted snow demons. My fortune had said, You are about to cross the great waters. But how, now, to begin?" And the sentence from the first page that reads, "I had a vision of dirt and rocks being poured over my chest," has been changed to "She had a vision of dirt and rocks being poured over my chest."

Should the reader too quickly dismiss the work either as yet another example of a random text generator or on the basis of its apparent arbitrary structure or its unreadability, it is important to note that using the links set by the authors, the piece always begins on the first page and proceeds methodically from one page to the next. That is, with only one mouse-over on each page, the text can change in only one predetermined manner at a time. Whereas procedural works such as Raymond Queneau's *Cent Mille Milliards de Poèmes* may give the impression of bearing only arbitrarily constructed meaning(s), this work allows for readerly intervention at the same time it foregrounds its constructedness or the way in which it is written to be read in one particular manner. Also, the order of the text becomes random only when the reader clicks on the small square at the top right of the screen and is then taken to whatever page number has been typed into the box. Ironically, only when the reader uses the computer-simulated page turner does the text become nonlinear and unstable. Not surprisingly, the pages from "The Jew's Daughter" are resolutely of the digital medium. They can be neither printed out nor cut and pasted to facilitate an

Will she disappear? That day has passed like any other. I said to you, "Be careful. Today is a strange day" and that was the end of it. I had written impassioned letters that expressed the urgency of my situation. I wrote to you that that it would not be forgivable, that it would be a violation of our exchange, in fact, a criminal negligence were I to fail to come through. To hand to you the consecrated sum of your gifts, the secret you imparted persistently and without knowledge, these expressions of your will that lured, and, in a cumulative fashion, became a message. In any case, the way things worked. Incorrigible. Stops and starts, overburdened nerves, cowardice (Is this what they said?), inadequacy, and, as a last resort, an inexplicable refusal. You asked could I build you from a pile of anonymous limbs and parts. I rarely slept and repeatedly during the night, when the moon was in my window, I had a vision of dirt and rocks being poured over my chest by the silver spade of a shovel. And then I would wake up with everything. It was all there like icons contained in a sphere and beginning to fuse together. When I tried to look at it, my eyes burned until I could almost see it in the room like a spectral yellow fire.

A street, a house, a room.

close

Will she disappear? That day has passed like any other. I said to you, "Be careful. Today is a strange day" and that was the end of it. I had written impassioned letters that expressed the urgency of my situation. I wrote to you that that it would not be forgivable, that it would be a violation of our exchange, in fact, a criminal negligence were I to fail to come through. To hand to you the consecrated sum of your gifts, the secret you imparted persistently. June through clouds like sculpted snow demons. My fortune had said, You are about to cross the great waters. But how, now, to begin? After stops and starts, overburdened nerves, cowardice, inadequacy, inexplicable refusal, after everything, she is still here, dreaming just outside the door, her affirming flesh beached in bed as the windows begin to turn blue. And what can now be said about this sleeping remainder? Her face is a pale round moon. She had a vision of dirt and rocks being poured over my chest by the silver spade of a shovel. And then I would wake up with everything. It was all there like icons contained in a sphere and beginning to fuse together. When I tried to look at it, my eyes burned until I could almost see it in the room like a spectral yellow fire.

A street, a house, a room.

close

FIGURES 34A AND 34B. *Screenshot of the first and second pages of Judd Morrissey's "The Jew's Daughter" from 2000.*

immobilization of the text for its scrutiny or to bring to bear techniques of close reading that apply only to the bookbound. (It is also likely that since we cannot print it out, this 608-page text will never be read in its entirety—thereby further setting itself apart from bookbound conventions of reading/writing narrative.)

Like Dickinson's manuscript poems, which draw our attention to both the limits and the possibilities of the paper-and-pen interface, as well as the singleness/doubleness of semantic meaning, "The Jew's Daughter" builds on a Dickinsonian critique as its mediation through the digital computer works against both easy assumptions about the linearity/nonlinearity of the page—even as it emulates the page—and the increasing transparency of the structure and the function of the hyperlinks—again, even as it emulates the conventional appearance of the link. It gestures toward markers of familiarity and legibility at the same time it undoes these same markers. As Morrissey tellingly put it in an interview with Matthew Mirapaul that appeared in a July 2000 article of the *New York Times*, "Because it takes the paradigm of the page, you can see that it's not a page."[53] Could we not say the same of a handwritten manuscript poem by Dickinson?

The Googlization of Literature

> Having a computer write poems for you is old hat. What's
> new is that . . . writers are now exploiting the language-
> based search engines and social networking sites as source
> text. Having a stand-alone program that can generate
> whimsical poems on your computer feels quaint compared
> to the spew of the massive word generators out there on the
> Web, tapping into our collective mind.
>
> —Kenneth Goldsmith, *Uncreative Writing*

> We are not Google's customers: we are its product. We—our
> fancies, fetishes, predilections, and preferences—are what
> Google sells to advertisers. When we use Google to find out
> things on the Web, Google uses our Web searches to find
> out things about us.
>
> —Siva Vaidhyanathan, *The Googlization of Everything*

Readingwriting

Throughout this book I have attempted to create a friction
between new and old writing interfaces while describing the
media poetics of writers themselves reading, through writing,
writing interfaces. Now that we are all constantly connected to
networks, driven by invisible, formidable algorithms, the role
of the writer and the nature of writing itself is being significantly transformed. Media poetics is fast becoming a practice
not just of experimenting with the limits and the possibilities
of writing interfaces but rather of *readingwriting*—the practice
of writing through the network, which as it tracks, indexes, and
algorithmizes every click and every bit of text we enter into
the network is itself constantly reading our writing and writing
our reading. This strange blurring of and even feedback loop

between reading and writing signals a definitive shift in the nature and the definition of literature.

Drawing on media archaeology—perhaps more freely, more creatively—as a means to analyze not only past/present media but also the past/present literary practices explicitly dependent on and, even, exploitative of media, we can look back from the vantage of the present to see that poets have been writing with the aid of digital computer algorithms since Max Bense and Theo Lutz first experimented with computer-generated writing in 1959 (though it is writing of a sort not familiar to us, writing as input and writing as choosing). Those early works are digital poems just as much as anything now called a digital poem or digital literature and produced with, granted, substantially more complex algorithms. What *is* new and particular to the twenty-first-century literary landscape is a revived interest in the underlying workings of the algorithms that are reading, writing, and reading our writing. Clearly aligned with the different incarnations of media poetics, tinkering, and the exploration of meaning-as-making, writers such as Bill Kennedy, Darren Wershler, Tan Lin, and John Cayley/Daniel C. Howe are concerned not just with the surface-level effects and results that characterized much of the fascination with computer-generated writing in the 1970s and the 1980s but with the ever-increasing power of algorithms—especially, search engine algorithms that attempt to know us, to anticipate and so shape our desires—and they work against the grain of that other seemingly user-friendly, invisible interface, Google's search engine. As Google itself put it prior to releasing their terms of service in March 2012, clearly echoing many of Apple's favorite marketing slogans, their aim is to "create a beautifully simple, intuitive user experience across Google."[1] Challenging us to resist the seductive pull of these simple, supposedly intuitive user interfaces, writers' creative misuse of Google prompts us to see that a passive acceptance of these algorithms necessarily means we cannot have a sense of the shape and the scope of how they determine our access to information, let alone shape our sense of self, which is increasingly driven by autocomplete, autocorrect, automata. It is a twenty-first-century network-based literary realization of Félix Guattari's declaration, in the wake of the failed uprising in Paris in May 1968, that "it's better to have ten consecutive failures or insignificant results than a besotted passivity before the mechanisms of retrieval."[2]

By the time John Battelle had published the first critical study on Google and the culture of search in 2005, *The Search: How Google and Its Rivals Rewrote the Culture of Business and Transformed Our Culture,* this search engine had so saturated our culture that Battelle could confidently write that Google was the "de facto interface for computing in the information age."[3] With the impending launch of Google Glass, Google could be poised to become the next Apple—and end an era inaugurated in 1984 with the release of the Macintosh. Early on, Battelle sensed this move away from the model of computing spawned from the Mac. He writes that once he had seen "Google's Zeitgeist,"

> I knew my beloved Macintosh had been trumped. Every day, millions upon millions of people lean forward into their computer screens and pour their wants, fears, and intentions into the simple colors and brilliant white background of Google.com. . . . "Toxic EPA Westchester County," a potential homeowner might ask, speaking in the increasingly ubiquitous, sophisticated, and evolving grammar of the Google search keyword.[4]

Google Glass—worn as a pair of eyeglasses—wants to be the opposite of something as noticeable and banal as eyeglasses. In many ways, it is the logical extension of devices such as the iPad, as it appropriates the same design principles of seamlessness, simplicity, and a desire for near invisibility and pares them down to literal invisibility while rendering Google-mediated information itself ubiquitous through a constant ambient stream of information that's activated through folksy commands such as, "OK Glass, Google photos of [search query]," and, "OK Glass, [question]?" Glass may even provide, in Google's own words, *"Answers without having to ask."*[5]

The "Googlization of literature" describes a collection of unique contributions to contemporary poetry, poetics, and even media studies: works of readingwriting that explore a twenty-first-century media poetics that questions how search engines answer our questions (whether we ask them or not), how they read our writing, and even how they write for us. Building on the twentieth century's computer-generated texts, these works of readingwriting give us a poetics perfectly appropriate for our current cultural moment in that they implicitly acknowledge we are living not just in an era of the search engine algorithm but in an era of what Siva Vaidhyanathan calls "The Googlization of Everything." When we search for data on the Web, we are no longer "searching"—instead, we are "Googling."[6] But reading-writers—in this case, conceptual writers turned digital writers or, in the case of John Cayley, digital writers turned conceptual writers via book publication—who experiment with/on Google are not simply pointing to its ubiquity but also implicitly questioning how it works, how it generates the results it does, and so how it sells ourselves and our language back to us.

The impetus of this literary critique of Google is aligned with that of early works of Internet art such as the *Web Stalker* from 1997—an experimental Web browser or piece of "speculative software" created by the art collective I/O/D (consisting of Simon Pope, Colin Green, and Matthew Fuller).[7] *Web Stalker* essentially turns the Web inside out, presenting the viewer/navigator with the HTML code of a given page and a visualization of all links leading to and from the page (see Figure 35). Similar to the works of media poetics I have discussed, *Web Stalker* is an artistic tool for drawing attention to the limits and the possibilities of a particular reading/writing interface, the Web browser. As cocreator Colin Green put it in a 1998 interview with Geert Lovink, "Browsers made by the two best-known players frame most peoples' experience of the web.

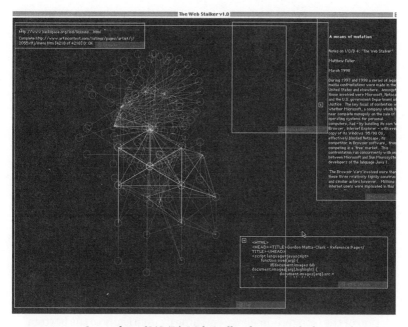

FIGURE 35. *Screenshot of I/O/D's* Web Stalker *browser, which was released in 1997.*

This is a literal framing. Whatever happens within the window of Explorer, for instance, is the limit of possibility."[8] The foregoing is then followed by Matthew Fuller's clarification that *Web Stalker* "is not setting itself up as a universal device, a proprietary switching system for the general intelligence, but a sensorium—a mode of sensing, knowing and doing on the web that makes its propensities—and as importantly, some at least of those 'of the web' that were hitherto hidden—clear."

Now, several generations of Internet culture later, not only are our choices of Web browser as limited and limiting as they were in the late 1990s, but Google has a monopoly on the the layer of information beyond the browser. As Lovink succinctly puts it fourteen years after interviewing I/O/D: "Google actively undermines the autonomy of the PC as a universal computational device.... The majority of users ... are happily abandoning the power to self-govern their informational resources."[9] Thus, in response, Internet art of the present moment takes the form of works like *Google Will Eat Itself* (*GWEI*).[10] A project of Ubermorgen.com (Alessandro Ludovico and Paulo Cirio), *GWEI* generates money by displaying Google advertisements on a network of websites hidden from search engine crawlers. The duo then uses the money from clicks on these ads to buy Google shares so that, as they put it, "We buy Google via their own advertisement! Google eats itself—but in the end 'we' own it!"[11] The final step in the *GWEI* process is that Google shares are then given to the Google To The People Public Company, which then distributes the shares back to the users who first clicked on the ads to initiate *GWEI*. It is an insurgent tactic that attempts to dismantle (from within the search engine itself) click-based advertising and surveillance on the Web in order to turn over ownership of information to the general public.

Google Gravity, by Ricardo Cabello Miguel, creates an animation of the Google homepage crashing dramatically to the bottom of the screen, along with anything you search for through the *Google Gravity* interface—cleverly turning the unprovoked, undesired browser crash usually driven by a glitch in the underlying code into a celebratory crash driven by perfectly functioning underlying code (see Figure 36).[12]

Constant Dullart's *The Revolving Internet,* from 2010, turns the Google homepage into a revolving windmill to the tune of Dusty Springfield's "The Windmills of Your Mind." The work turns into an elegy for the end of a love affair with Google and its voracious appetite for archiving traces of material culture and quickly transforming these traces into memories: "Pictures hanging in a hallway / And a fragment of this song / Half remembered names and faces / But to whom do they belong? / When you knew that it was over / Were you suddenly aware / That the autumn leaves were turning / To the color of her hair" (see Figure 37).[13]

In 2012, likely coinciding with Google's release of a new streamlined terms of service (TOS) that consolidated seventy different TOS for different Google products, such as Gmail and

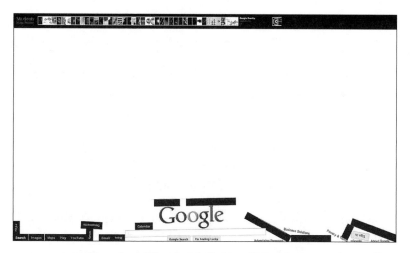

FIGURE 36. *Screenshot of Ricardo Cabello Miguel's* Google Gravity *from 2009.*

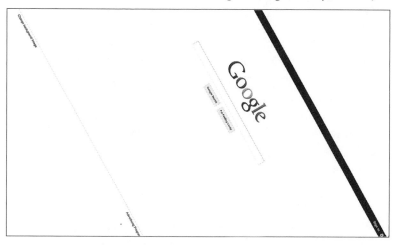

FIGURE 37. *Screenshot of Constant Dullart's* The Revolving Internet *from 2010.*

YouTube, into one TOS, Constant Dullart followed *The Revolving Internet* with *Terms of Service*. This work of Internet art quite simply turns the Google search bar into an animation of a mouth reciting the astonishingly long and alarming terms for using Google products (which is quickly amounting to terms for using the Web). We find, for instance, that "Google may modify the Terms of Use at any time with or without notice," and if this now standard wording for a TOS is not cause for worry, we then find that this work by Constant Dullart (along with *The Revolving Internet*) clearly violates—and enacts this violation of—Google's TOS, which explicitly states we are forbidden to "misuse" and "interfere with" Google's services.[14] Of course, this wouldn't be exceptional if it weren't for the fact that Google was, again, our de facto interface for accessing information online. Thus, our agreement to these TOS amounts to our consenting to passive consumption of whatever information Google chooses to present to us in whatever form it deems appropriate (for its commercial interests, not the public good).

Whereas the digital seems to quickly erode distinctions between genres and art practices, works of readingwriting not only are interested in the broad political implications of the googlization of everything but also—given that they emerge from a lineage more literary than Internet art's—are interested in the effect of googlization on (or just the algorithmically driven commodification of) our language, on what and how we read and write. These writers seek to acknowledge a materiality of language in the digital that goes deeper than an acknowledgement of the material size, shape, sound, and texture of letters and words that characterize much of twentieth-century bookbound, experimental poetry practices. They take us beyond the twentieth-century avant-garde's interest in the verbal/vocal/visual aspect of materiality and instead urge us to attend to the materiality of twenty-first-century digital-language production. They ask, What happens when we appropriate the role of Google for purposes of reading/writing other than Google's? What happens when we wrest Google from itself and instead use it not only to find out things about us as a culture but to read and write what Google is finding out about us? What happens when we subvert its single-minded drive to turn the Web into what Evgeny Morozov calls a world of "frictionless, continuous shopping"—a world in which Google's search algorithms do not define our access to information so much as, through programs like Shopping Express and autonomous search, steer us toward purchasing products, a world "where we no longer need to search for anything, since we ourselves are perpetually monitored, with the relevant product or information sent to us based on perceived need"?[15]

This cluster of readingwriting that both probes and is driven by the search engine enacts a study of software. Lev Manovich writes in *Software Takes Command,* "Software Studies has to investigate both the role of software in forming contemporary culture, and cultural, social, and economic forces that are shaping development of software itself."[16] If the search engine is currently one of the most powerful pieces of cultural software, then this literary critique of Google positions itself as a mode of twenty-first-century media poetics. Framed as such, even though work by readingwriting authors such as Darren Wershler and Bill Kennedy is usually referred to as "conceptual writing," it is a distinct departure from the twentieth-century dictums of conceptual art such as the following by Sol Lewitt, which is frequently cited as an explanation and, perhaps even, a justification of conceptual writing:

> In conceptual art the idea or concept is the most important aspect of the work. When an artist uses a conceptual form of art, it means that all of the planning and decisions are made beforehand and the execution is a perfunctory affair. The idea becomes a machine that makes the art.[17]

While it has never been framed as conceptual writing, much of the computer-generated poetry produced with stand-alone software prior to the advent of the Web exemplifies a poetry that takes up Lewitt's urging to develop an approach that is invested in the idea of producing text with a computer rather than in attending to what is produced—such that "the execution is a perfunctory affair." My argument is, however, that twenty-first-century readingwriting ups the ante. It is intimately involved in both planning the execution of the work and directing and even critiquing the execution itself—in this case, using Google and thus critiquing Google—often expressed by way of the output/outsourced text.

Computer-Generated Writing and the Neutrality of the Machine

If we look briefly at how the discourse on and of text generation gradually shifts from the late 1960s to the mid-2000s, we can trace an evolutionary line that leads to experiments with the digital computer algorithm in literary practices, the way in which the algorithm has gained cultural dominance via Google, and the way in which it has receded from view.

One of the earliest published collections of computer-generated writing appears in the 1968 *Cybernetic Serendipity,* which catalogs an exhibit by the same name curated by Jasia Reichardt.[18] As the entire exhibit was deeply concerned with the role of the algorithm—defined in the introductory glossary of terms as "a prescribed set of well-defined rules for the solution of a problem"—the part of the collection dedicated to "computer poems and texts" is noteworthy for two reasons.[19] First, every example of computer-generated text is accompanied by a detailed explanation or illustration of the "well-defined rules" that produced the text—from the method of input, the programming language, and the computer used to the vocabulary and so-called semantic schema from which poems or sentences were generated (see Figure 38). This selection of computer-generated texts in *Cybernetic Serendipity* is, with just a few exceptions, among the first and the last to lay bare its underlying mechanisms and thus advocate for a thoroughgoing understanding of how the production process necessarily impacts the written product. (The foregoing is likely due in part to the gradual shift that took place from the late 1960s to the early 1980s from tinkering with computers as open-source circuit

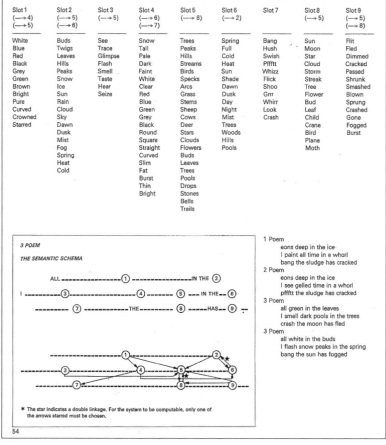

FIGURE 38. *"Computerized Japanese Haiku,"* the schema for a haiku-generating program written in TRAC with poems produced by Margaret Masterman and Robin McKinnon Wood and published in Cybernetic Serendipity.

boards for hobbyists to mass producing computers as closed, prepackaged, supposedly user-friendly devices.)

Even Richard Bailey's 1973 *Computer Poems*—which opens with the declaration that "computer poetry is warfare carried out by other means, a warfare against conventionality and language that has become automatized"—contains only the textual results of computer generation and no explanation of how these texts were produced.[20] The closest we get to an exploration of process is the work "TIMESHARING" by Archie Donald, yet even here, Donald adopts the syntax of the programming language BASIC to write a faux code poem (see Figure 39).[21]

One could argue that *Computer Poems* was one of the more obscure collections of computer-generated poetry and so might not be particularly indicative of anything. However, even the famed 1984 collection of poems *The Policeman's Beard Is Half-Constructed,* by the artificial intelligence program Racter, neatly avoids any discussion of the specifics of the production process (possibly to preemptively draw our attention away from questions about how these poems were written, as it was later discovered that substantial human intervention was involved in crafting them).[22] As Racter cocreator William Chamberlain puts it: "The specifics of the communication in this instance would prove of less importance than the fact that the computer was in fact communicating something. In other words, what the computer says would be secondary to the fact that it says it correctly. Computers are supposed to compute. . . . *They are tools we employ to get certain jobs done.*"[23]

Charles O. Hartman and Hugh Kenner's *Sentences* from 1995 includes a lengthy afterword in which Kenner does explain the workings of the Travesty and DIASTEXT programs used to generate sentences based on the nineteenth-century grammar-school book *Sentences for Analysis and Parsing.* Yet the overarching point Kenner wants us to attend to is that even when we use an algorithm to generate text, we ought not to "underrate our contributions." He assures us there is still ample authorial intent driving the shape of the final product, and so

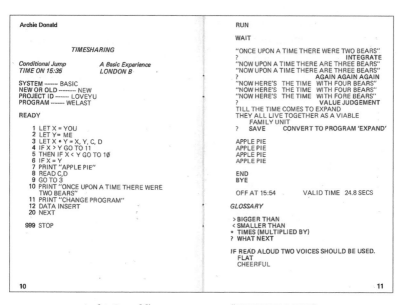

FIGURE 39. *Archie Donald's computer poem "TIMESHARING" as it appears in Richard Bailey's 1973 collection* Computer Poems.

we need not worry too much about the underlying processes determining what can and cannot be produced.[24] Twenty-two years after Bailey's *Computer Poems*, ultimately, those who are invested in computer generation are likewise still invested in the idea of a single author who produces texts that are recognizably poems or recognizably literary. The ways in which the software, even the computer itself, are ideologically driven (and so anything but neutral) are still taken for granted or glossed over altogether.

That said, the most recent and notorious feat of computer-generated writing does directly challenge us to recognize how outdated this habitual belief is in the ideology of the single, intending original author—one that is unquestionably unsettled in the age of the digital computer algorithm and not simply shifted, as Kenner would have us believe. Stephen McLaughlin and Jim Carpenter's fall 2008 release of the first (and only) issue of the poetry journal *Issue 1* is a 3,785-page-long collection of computer-generated poems attributed to over 3,164 poets who of course did not write a single poem in the collection.[25] However playfully pointed their gesture, *Issue 1* was quickly met with virulent hostility, even by poets long associated with challenging equally outdated notions that poems ought to be instances of supposedly clear, direct, emotion-filled self-expression. The most surprising expression of outrage came from Language poet Ron Silliman, who called the collection an act of "anarcho-flarf vandalism" before issuing an indirect order to cease and desist:

> I might note that the last time I felt ripped off by an on-line stunt, I sued—as a lead plaintiff in a class-action case brought by the National Writers Union. And while I can't discuss the suit, as a condition of the subsequent settlement, I will note that we could have gotten a pretty good major league middle infielder for the final amount. Play with other people's reps at your own risk.[26]

Suddenly and clearly, we see not only that the singular, intending author is alive and well, even in experimental poetry circles, but that—with no acknowledgment or even curiosity about the programming feat that produced such a massive amount of passably good poems—so too is the belief in the overriding value of poetic product over poetic process (and this despite the perpetual alignment of Language poetry with an investment in the materiality of the word).

Thus, to return to *Cybernetic Serendipity*, despite its interest in laying bare the underlying mechanisms for text generation (an interest that was soon abandoned in the 1970s), this collection seems to have inaugurated an era of literary text production driven by a belief in the "neutrality of the machine"—a belief that could also account for the gradual waning in subsequent decades of interest in making computer processes transparent. A statement by Marc Adrian opens the section on "computer poems and texts" with the assertion: "To me the neutrality of the machine is of great importance. . . . It allows the spectator to find his own meanings in the association of words more easily, since their choice, size and disposition are determined at random."[27] No doubt because we are gradually becoming ever more aware of the absence of the random in search algorithms, as well as their power not just to determine our access to information but even to *predetermine* our likes and dislikes (those traditional markers of how we assert our identity), this era of the "neutrality of the machine" now seems to be slowly shifting.

"And So They Came to Inhabit the Realm of the Very Unimaginary"

This assumption that tools are inherently neutral, neither-good nor bad, is precisely what separates twentieth-century computer-generated writing from twenty-first-century reading-writing. The latter has internalized one of the basic tenets of media study, as put forward by Marshall McLuhan in his 1964 "The Medium Is the Message," that (writing) media, that machines, are profoundly ideological and therefore anything but neutral.[28] Further, we cannot hope to understand the way in which a machine shapes us and our experiences by studying only surface effects or, in this case, the text that a computer produces and not the production process itself. While computer-generated writing and readingwriting share a common tool for creating texts—the algorithm—one of the latter's critical innovations is that it not only frames the how and the why of works that depend upon the algorithm underlying any given search engine but also foregrounds its own constructedness as a way of making visible the invisible, taken-for-granted media that delimit what information we can and cannot access.[29]

If we attend to the underlying mechanical differences between computer-generated poetry and poetry that is the result of a search engine query, then the primary difference between the two is that in the former the poet or programmer creates the data set, as well as the grammatical and syntactical parameters, from which the computer creates a poem and that in the latter the poet initiates a search in order to obtain a data set from which to draw. Searching is itself an act of curation, as one must judiciously choose in order to obtain any meaningful results—or rather, searching reframes writing as choosing or arranging. Since most computer-generated poems are often the result of judicious choosing among phrases, sentences, entire poems, the difference that makes a difference between the two poetry practices is not what the poet/programmer does with the data set but rather the nature of the data set itself. The difference between even the most expansive set of data underlying a computer-generated poem and one that comes from Google is the difference between one person's or even one set of people's language practices/preferences and the language practices/preferences of every primarily English-speaking culture, both past and present, that has access to the Web.

One of the most compelling examples of readingwriting that uses Google to read our culture and even Google itself is Bill Kennedy and Darren Wershler's *apostrophe*.[30] As they explain in their afterword, *apostrophe* began as a list poem of "you are" statements that Kennedy wrote in 1993. Kennedy and Wershler then built the Web-based "apostrophe engine," which searched (first on AltaVista in April 2001 and later on Google) for words from Kennedy's original poem and then picked out all the "you are" statements from the results to create a poem written by "us." The final step in their experiment in media poetics turned to the medium of the book as they ran searches from September to October 2002 and subsequently published this set of search results, along with a handful of search results from AltaVista the previous year, as *apostrophe* in 2006.[31]

Given that the apostrophe engine is a search engine that searches particular search engines to create poems that are the bases for yet more searches of search engines, the apostrophe engine is a meta–search engine that provides us with bits of material evidence that reveal the ever more sophisticated workings of Google's search algorithm through the shape, the content, and the syntactical structure of the statements themselves. In 2001, for example, a search on AltaVista for the phrase "a home by the sea" produced, on the whole, lengthy statements that seemed to reflect individual attempts to express, define, and develop individual, complex identities: "you are a fool to waste your time reading any further • you are smart enough to do it now and you will, I promise, be one step closer to your dream than if you don't • you are invited to write it down simply because writing it down is a shortcut."[32] In 2002 phrases from Kennedy's original poem, such as "used and abused" and "a foreign agent who accidentally ruptured an emergency cyanide tooth cap," did not return any results on Google, but when Kennedy and Wershler re-searched the phrases in 2006, they did. The results indicated three key trends: (1) Google's more sophisticated system of search, (2) our own willingness to pour our everyday experiences online, and (3) the way in which Google archives had shifted in language usage and even driven the corporatization of language. In 2002, "we" wrote, "You are not comfortable with formal terms of logic, so it's best to stay away from this phrase, or risk embarrassing yourself," whereas in 2006, "our" writing shifted to the assertion, "You are the psychotic individual who placed the call, let us know and we'll send you a complimentary Circlemakers T-shirt!"[33] "You" stands in, then, for both our culture at large (for we don't need to make any of these assertions or even relate to them personally in order to read them broadly as signs of culture) and Google and the way in which it insistently mediates and drives the cultural "you." As such, the apostrophe engine perfectly enacts the invocation in the opening epigraph of poet Steve Venright's *Spiral Agitator*: "Build an engine with words. Let it make you speak." We may not have access to the source code of the apostrophe engine, but the way in which it "speaks" us is the way in which it critiques and, even, gives us the means by which to critique for ourselves the supposed neutrality of the Google search algorithm.

The second compelling example of the googlization of readingwriting is Tan Lin's 2008 *HEATH,* as it's titled on the spine of the book, or *plagiarism/outsource, notes towards the definition of culture, untitled Heath Ledger project, a history of the search engine, disco OS,* as it's titled on the front cover.[34] Just by looking at the front cover, the spine, and the back cover, we see that Lin approaches the centrality of the search engine less through the construction of a meta–search engine and more through a focus on a constellation of reading/writing software that includes the algorithmically driven search engine, along with autocorrect (hence, the seemingly incorrect yet oddly appropriate "untilted" in the title) and Microsoft Word. The design of the front and the back covers becomes more pointedly meaningful once we read "A NOTE ON THE DESIGN" by Danielle Aubert, in which she states, "[*HEATH*] was art directed by Danielle Aubert and designed by Tan Lin in Microsoft Word. The text is set in Courier except where text was imported directly from the Internet, in which case the original formatting is preserved."[35] The issue of how one separates generated writing from supposedly original writing is unsettled, as the text on the front cover is set in Courier (incidentally, a monospace font originally designed not for computers but for typewriters, and so its appearance throughout the book adds to Lin's media remix) and is wrapped so that "outsource" is split into "outsou" and "rce," suggesting an imbedded formatting that may have indeed been cut and pasted or plagiarized from the Web. Of course, "outsource" is itself a nod to the way in which this mode of readingwriting relies on other sources, located elsewhere, to do the labor of providing the linguistic data set from which Lin writes by way of selection. The writer as one who produces original work shifts to one who chooses what and how to copy.

Turning to the back cover, we find a list of eight names— seven of which belong to students in Lin's Asian American writer's workshop and whose handwritten autobiographies are included, as Danny Snelson points out, "as outsourced bibliographic production."[36] The students are authors quite alongside Lin, which is appropriate given the enlarged copyleft symbol on the back cover that hovers above an equally enlarged feed icon. If copyleft is the means by which one makes "a program (or other work) free, and [requires] all modified and extended versions of the program to be free as well," then *HEATH*—published under the Creative Commons Attribution–Share Alike 3.0 Unported License—is a book-machine whose content comes from outsourced writing, which in turn opens the book for others to pillage for their own outsourced writing experiments.[37] As "he" or "they" or "we" write roughly midway through the book, "As of this writing, the copyleft symbol has no legal status in the United States, but its attachment to this work is meant to facilitate, by offering a non-legal license, to other users to copy and redistribute this material."[38] If a feed icon represents the act of a user/reader initiating a subscription to new content, then *HEATH* is new content that resides between the realm of the book and the realm of the digital at the same time it, like *apostrophe,* is a metawork—a work about the larger network within which it is nested and upon which it depends. As Lin puts it, "'This' work is Nominally a novel inside a Network"—gesturing with "this" to the impossibility of defining firm boundaries around any text that is always already a part of a network.[39]

Whereas *apostrophe* is firmly engaged with the underlying workings of one particular network—the search engine—the search engine is only one of many networks represented in *HEATH,* though it is positioned as the most crucial. At nearly every turn, Lin pushes to the foreground the ubiquitous, invisible, and usually unquestioned presence of Google and the ways in which its algorithms underlie our everyday reading/writing practices. Both his acts of plagiarism—such as his inclusion of substantial portions of the e-text of *The Diary of Samuel Pepys,* articles from *Critical Inquiry,* excerpts from Google Books, and online ads for Blimpie—and his anecdotes of everyday conversations, such as one from a dinner partner who talks "about doing a Likeness Search™ for Mischa Barton's Darling Long Drop Earrings," are entirely mediated by Google, as evidenced by Lin's inclusion of screenshots from searches or of the Google Toolbar for Safari and Firefox. He seems to understand perfectly the way in which the search engine permeates our worldview such that there is no perceivable boundary between the online world and the "real" world—a supposed boundary that has been used to justify the primacy of the one over the other and so the primacy of bookbound values, tied up with notions of the singular author and the importance of originality, over digital values. For example, Lin writes early on in a semi-autobiographical mode: "And so they came to inhabit the realm of the very unimaginary, for most of them the Pickwick Arms, with its faded armchairs, dirty carpet . . . it comprised a factory of miscellaneous bliss, a search engine not of emotions but of the most unformed and standardized of affects, personal productivity software, etc." He continues some pages later: "Their eyes were made of search engines like a search engine."[40] Arguably, Lin's primary motive in placing Google at the center of the production process for *HEATH* is more to observe the googlization of everything than to enact the kind of critique at the heart of *apostrophe*. In other words, Lin's interest seems to lie more in bringing to the fore his own readingwriting practice at the level of a user/consumer of Google's search engine than in drawing attention to the constructedness of the search engine.

Still, writing through or of our search engine–driven culture in *HEATH* helps to make transparent the production process behind it, the reading/writing software used, its mediation through Google, and the way in which this network of reading/writing software (pre)determines the visual and semantic shape of *HEATH*, all of which ideally facilitates others' plagiarism of his work. That said, what both *HEATH* and *apostrophe* share is a determination to bring the literary well into the twenty-first century through a media poetics that advocates for *reproduction* via knowledge of *production*.[41]

Finally and in contrast with Lin's work, John Cayley and Daniel C. Howe's *How It Is in Common Tongues* is intensely concerned with drawing attention to the profound influence of Google's search engine and how it works on readingwriting practices.[42] Taken up as part of the larger *The Reader's Project* began by Cayley and Howe in 2009 and as part of another subproject of *The Reader's Project* called *Common Tongues, How It Is in Common Tongues* brings us full circle to the works of digital literature I discuss in chapter 1, which are a disruptive response to the computing industry's insistent drive to create devices that are nearly invisible. The work claims it is a "reading" of the English version of Samuel Beckett's *How It Is*, published in 1964, in that it uses this source text to generate new poetics texts via "a universally accessible search engine." The authors delineate their readingwriting process as follows at the end of their version of *How It Is*:

> This book was composed by searching for the text of Samuel
> Beckett's *How It Is* using a universally accessible search
> engine, attempting to find, in sequence, the longest com-
> mon phrases from *How It Is* that were composed by writers
> or writing machines other than Beckett. These phrases are
> quoted from a portion of the commons of language that
> happens to have been indexed by a universally accessible
> engine.[43]

Thus, Cayley and Howe's *How It Is in Common Tongues,* subtitled "Cited from the Commons of digitally inscribed writing" and reading on the bibliography page "Copyleft, 2012" by "The Natural Language Liberation Front," appears as nothing but pages and pages of words and phrases footnoted with URLs discovered via a search engine.

What is not stated explicitly is that the "universally accessible search engine" the authors used was Google's and, more, that the work attempts to address the way in which this search engine is quickly commodifying "our linguistic cultural commons."[44] *How*

It Is in Common Tongues is made entirely of the words of others yet is also utterly mediated by Google's search engine algorithm, which predetermines the text supplied by the commons we can see or access. In contrast to earlier writers' belief in the neutrality of the machine, Cayley and Howe understand that algorithms are not simple, straightforward purveyors of information. More, it is not coincidental that these two long-standing members of the e-literature, digital literature, and digital poetry communities chose to take on this practice of readingwriting and then publish their results in a print-on-demand book—for that supposedly antiquated device, the book, is fast becoming a safe haven for readingwriting because its particulars cannot be tracked, monitored, indexed, fed into an algorithm, and given back to us as a commodity.

Perhaps, the future of digital literature is readingwriting that is born of the network but lives offline—digital literature transformed into bookbound readingwriting that performs and embodies its own frictional media archaeological analysis.

Notes

Introduction

    1. Florian Cramer, "What Is Interface Aesthetics, Or What Could It Be (Not)?," in *Interface Criticism: Aesthetics beyond Buttons*, ed. Christian Ulrik Andersen and Sren Pold (Aarhus, Denmark: Aarhus University Press), 119.

    2. Johanna Drucker, "Humanities Approaches to Interface Theory," *Culture Machine* 12 (2011): 10. Drucker echoes this expansive definition of *interface*, writing that "a book is an interface, so is a newspaper page, a bathroom faucet, a car dashboard, an ATM machine."

    3. Alexander R. Galloway, "The Unworkable Interface," *New Literary History* 39, no. 4 (Autumn 2008): 936.

    4. Janet Murray, *Inventing the Medium: Principles of Interaction Design as a Cultural Practice* (Cambridge, Mass.: MIT Press, 2011), 10.

    5. Alexander R. Galloway, *The Interface Effect* (Malden, Mass.: Polity, 2012), 25.

    6. Friedrich Kittler, *Gramophone, Film, Typewriter*, trans. Geoffrey Winthrop-Young (Stanford, Calif.: Stanford University Press, 1999), xxxix.

    7. Geert Lovink, *My First Recession* (Rotterdam: V2/NAi Publishers, 2003), 11.

    8. Michel Foucault, *The Archaeology of Knowledge*, trans. A. M. Sheridan Smith (New York: Random House, 1972). That said, media archaeology is a (deliberately) frustrating field because it does not have a clear, overall methodology with precise parameters and a driving philosophy. It does find its roots in Michel Foucault's *The Archaeology of Knowledge*, in which Foucault thinks through the archive as a system that governs discourse. The field also finds roots in work by Marshall McLuhan that's roughly contemporaneous to that by Foucault, as well as in Kittler's analysis of discourse networks. Beyond these three thinkers, however, the field splinters into those who have an interest in reviving Foucault's notion of the archive in a digital context, those who want to inaugurate a new way to think about (media) history, those who want to renovate dead media or imaginary media, those who are looking for a theoretical framework by which to look at the particular material dimensions of machines, and many other variations. Furthermore, work on digital

forensics by Matthew Kirschenbaum and on media history by Lisa Gitelman is not explicitly identified with media archaeology but bears undeniably implicit ties to the field and also provides crucial theoretical and methodological frameworks that can be extended more specifically to interfaces.

9. Media Archaeology Lab, http://mediaarchaeologylab.com. There *are* a small handful of sibling organizations in the United States—such as the Maryland Institute for Technology in the Humanities' collection of vintage computers, Nick Montfort's Trope Tank at MIT, and Dene Grigar's early electronic literature lab at Washington State University. With the exception of the foregoing, the majority of vintage computer collections are more akin to archives or special collections than they are labs in the sense of being utterly open spaces for hands-on teaching and research.

10. Jim Andrews, "Framing 'Off-Screen Romance,'" http://vispo. com/bp/jim.htm.

1. Indistinguishable from Magic

1. Friedrich Kittler, "There Is No Software," *Ctheory*, October 18, 1995, http://www.ctheory.net/articles.aspx?id=74.

2. Rob Swigart, "A Writer's Desktop," in *The Art of Human-Computer Interface Design*, ed. Brenda Laurel (Reading, Mass.: Addison-Wesley Professional, 1990), 140–41.

3. Steven Johnson, *Interface Culture: How New Technology Transforms the Way We Create and Communicate* (New York: Basic Books, 1997), 213.

4. Fluid Interfaces Group, "Research Vision," http://fluid.media. mit.edu/vision.

5. Roel Vertegaal and Ivan Poupyrev, introduction to "Organic User Interfaces," ed. Roel Vertegaal and Ivan Poupyrev, special issue, *Communications of the ACM* 51, no. 6 (June 2008), http://www.organicui. org/?page_id=26.

6. Ibid.

7. Natural User Interface Group, "About the NUI Group," http:// nuigroup.com/log/about.

8. Daniel Saffer, *Designing for Interaction: Creating Innovative Applications and Devices*, 2nd ed. (Berkeley, Calif.: New Riders, 2010), 17.

9. Daniel Terdiman, "IBM: Mind Reading Is Less than Five Years Away. For Real," *Geek Gestalt* blog, CNET News, December 19, 2011, http:// news.cnet.com/8301-13772_3-57344881-52/ibm-mind-reading-is-less -than-five-years-away-for-real.

10. Adam Greenfield, *Everyware: The Dawning Age of Ubiquitous Computing* (Berkeley, Calif.: New Riders, 2006), 137–38.

11. Ben Schneiderman and Catherine Plaisant, *Designing the User Interface: Strategies for Effective Human-Computer Interaction,* 5th ed.(Upper Saddle River, N.J.: Pearson, 2010), 194.

12. Donald Norman, "Natural User Interfaces Are Not Natural," *Interactions,* May/June 2010, 6.

13. Saffer, *Designing for Interaction,* 216.

14. "Anand Agarawala: Rethink the Desktop with BumpTop," Ted.com video, 4:43, from a recording made March 2007, posted June 2007, http://www.ted.com/talks/anand_agarawala_demos_his_bumptop_ desktop.html; Fluid Interfaces Group, "Research Vision."

15. Mark Weiser and John Seeley Brown, "The Coming Age of Calm Technology," October 5, 1996, http://cs.ucsb.edu/~ebelding/courses/ 284/s06/papers/calm.pdf.

16. Ibid.; emphasis my own.

17. Mark Weiser and John Seeley Brown, "The World Is Not a Desktop," *Interactions,* January 1994, 8.

18. Ibid., 8.

19. Ibid., 8.

20. Mark Weiser, "The Computer for the 21st Century," *Scientific American,* September 1991, 94.

21. John Krumm, *Ubiquitous Computing Fundamentals* (Boca Raton, Fla.: Chapman and Hall, 2010), 5.

22. Weiser, "The Computer for the 21st Century," 99; emphasis my own.

23. Ibid., 99.

24. It's worth noting that ubicomp gets its name in part from Philip K. Dick's 1969 science fiction novel *Ubik.* No doubt, Weiser was originally making a playful, tongue-in-cheek reference to the mysterious yet ubiquitous product Ubik that weaves its way through the novel, perhaps acting as a metaphor for God or perhaps just as a sublime plot device that defies understanding. At one point, Dick's widow, Tessa Dick, wrote on her blog, "Ubik is a metaphor for God. Ubik is all-powerful and all-knowing, and Ubik is everywhere." If Tessa Dick's interpretation of Ubik reflects a common understanding of the fictional product, then it is ironic indeed, for despite claims that ubicomp was the next wave of computing after the desktop, even Weiser himself was not advocating for a system of computing that was all-powerful and all-knowing, effectively creating utterly passive, if not helpless, users. Philip K. Dick, *Ubik* (New York: Vintage Books, 1991); Tessa Dick, "Ubik Explained, Sort Of," *Tessa Dick Presents: It's a Philip K. Dick World!* blog, December 4, 2008, accessed February 12, 2010, http://tessadick.blogspot.com/2008/12/ ubik-explained-sort-of.html (page discontinued).

25. Tim Cook, "Tim Cook Unveils the New iPad Mini," YouTube video, 40:05, uploaded by Yung Jules, October 23, 2012, https://www.youtube.

com/watch?v=jXy7AhWofVM. There is a marked, long-standing tradition in computer advertising to align women or conventionally feminine qualities with the machines, one that is comparable to the history of using misogynist rhetoric to sell cars. In this way, the language used to sell the iPad is hardly unique—for example, Apple used fashion magazine–inspired language in 2008 for its iMac slogan "Beauty. Brains. And now more brawn"; its "You can't be too thin, Or too powerful" slogan in 2007 for the new iMac; and its "Small is Beautiful" slogan in 2006 for the new Intel Mac mini.

26. Ibid.

27. "Making an iPad This Small Was No Small Feat," Apple, accessed March 24, 2013, http://www.apple.com/ipad-mini/design (webpage since updated).

28. Steven Levy, *The Perfect Thing: How the iPod Shuffles Commerce, Culture, and Coolness* (New York: Simon & Schuster, 2006), 98; emphasis my own.

29. The principles of magic shows and sleight-of-hand techniques are at work even in the design of Apple stores, such as the flagship store in New York City, which has been made to appear as if it's within a glass cube (made of nonreflective glass to create an even more convincing illusion of a marvelous, even pure, reality) that sits above ground, when in fact the store is underneath. The glass cube is merely an elaborate ruse, and we, the customers, become complicit with the revealing of the trick's functioning as we descend the escalator and the actual store comes into view. This way, with every visit to the Apple store, we experience a revelation.

30. "Apple iPad: First Release Video of the iPad," YouTube video, 2:04, uploaded January 27, 2010, http://www.youtube.com/watch?v=HmCPcZk2Bgc.

31. Ibid.

32. Ibid.

33. Arthur C. Clarke, "Hazards of Prophecy: The Failure of Imagination," in *Profiles of the Future: An Enquiry into the Limits of the Possible* (New York: Harper & Row, 1962).

34. "Apple iPad: First Release Video of the iPad." While thinness was clearly an important selling point at the first iPad launch, just as it was for the release of the MacBook Air, which was marketed in 2008 using the phrases "The world's thinnest notebook" and "Thinnovation," it's mentioned only about five times, compared with the thirty-five times it was mentioned at the iPad mini launch. The iPad made it possible to "hold the internet in the palm of your hand." This oft-repeated phrase echoed the 2007 slogan for the iPhone, "The internet in your pocket," or even their 2001 slogan for the iPod, "1,000 songs in your pocket," and so became another "magical" feature of Apple devices. Not surprisingly,

since sleight of hand is often tied to a rhetoric of the natural and the invisible in interface design, the first iPad launch featured numerous exclamations like, "I don't have to change myself to fit the product. It fits me," and declarations about how the device was "intuitive, easy to use, fun to use, so [it] really fit the user."

35. Jesus Diaz, "The Next Step: iPad Is the Future," *Gizmodo* blog, April 2, 2010, http://gizmodo.com/5506692/ipad-is-the-future.

36. "iPad Commercial #2: iPad Is Magical," YouTube video, 0:30, uploaded by AppOSX, September 11, 2010, http://www.youtube.com/watch?v=8osEYSbA_Ps.

37. "Apple iPad TV Ad: iPad Is Delicious," YouTube video, 0:31, uploaded by rudis323, August 18, 2010, http://www.youtube.com/watch?v=btfbIVGES1I.

38. "iPad Is Electric," YouTube video, 0:29, uploaded by multiramagr, March 17, 2011, http://www.youtube.com/watch?v=cTwWi3Rx5s4.

39. "iPad Magic Sub:E," YouTube video, 2:45, uploaded by salarymagician, May 30, 2010, http://www.youtube.com/watch?v=ATpSPNIuj3M; " iPad Beer at Hofbraeuhaus by iSimon (iPad Zauberer)," YouTube video, 2:04, uploaded by iOSMagic, March 2, 2012, http://www.youtube.com/watch?feature=player_embedded&v=6a8Eimr-fmo. Matthew Kirschenbaum astutely reminded me of the now famous line from William Gibson's 1982 story "Burning Chrome" that "the street finds its own uses for things." In other words, these iPad magicians are not so much complicit in Apple's marketing as they represent the impressive, creative ingenuity of everyday users who find ways to work around and against the closed interface of the iPad, quite in spite of the formidable infrastructure that Apple has put in place precisely to prevent this sort of creative (mis)use.

40. Tim Cook, "Tim Cook Unveils the New iPad Mini"; Chris Foresman, "Apple Says First Weekend of 4G iPad, iPad Mini Sales Set Record," *Ars Technica* blog, November 5, 2012, http://arstechnica.com/apple/2012/11/apple-says-first-weekend-of-4g-ipad-ipad-mini-sales-set-record.

41. Walter Isaacson, *Steve Jobs* (New York: Simon & Schuster, 2011), 322; "What Are Creative Industries and Creative Economy," British Council website, http://creativecities.britishcouncil.org/creative-industries/what_are_creative_industries_and_creative_economy. At least since the early 2000s when Richard Florida published his massively successful *The Rise of the Creative Class,* entire countries have begun the push to co-opt creativity as a way to monetize the production of information and knowledge by supporting the growth of the so-called creative economy or creative industry. Or as the British Council puts it quite bluntly, the creative industry is "the framework through which creativity translates into economic value." While Apple is not unique in how it strategically invokes creativity as part of its marketing campaign, it was certainly one of the first—if not the first—major company to do so.

42. Myron K. Krueger, *Artificial Reality II,* 2nd ed. (Reading, Mass.: Addison-Wesley Professional, 1991), 64.

43. Myron K. Krueger, "Videoplace, Responsive Environment, 1972–1990s," YouTube video, 7:56, from a talk given June 10, 1988, uploaded by MediaArtTube, April 7, 2008, http://www.youtube.com/watch?v=dmmxVA5xhu0.

44. Krueger, *Artificial Reality,* xiv; Nick Montfort et al., *10 PRINT CHR$(205.5+RND(1));:GOTO 10* (Cambridge, Mass.: MIT Press, 2013), 165. Remarkably, in stark contrast to the way most universities now conceive of a liberal arts education, the writers of *10 PRINT CHR$ (205.5+RND(1));:GOTO 10* describe how "by 1971, 90 percent of the seven most recent classes of [Dartmouth] freshmen had received computer training," including learning how to program in BASIC.

45. Bill Buxton, "Multi-Touch Systems That I Have Known and Loved," last modified March 19, 2013, http://www.billbuxton.com/multitouchOverview.html.

46. Krueger, *Artificial Reality,* 37.

47. Ibid., 86; emphasis my own.

48. Ibid., 45.

49. Ibid., 51.

50. Ibid., 225.

51. Ibid., 226.

52. Apple Inc., "iOS Human Interface Guidelines," December 17, 2012, accessed March 24, 2013, http://developer.apple.com/library/ios/#documentation/UserExperience/Conceptual/MobileHIG/Introduction/Introduction.html (webpage has been since updated).

53. Ibid.

54. Joseph Bernstein, "Apple Rejects Game Based on Syrian Civil War," *Kill Screen,* January 7, 2013, http://killscreendaily.com/articles/news/apple-rejects-game-based-syrian-civil-war. *Endgame: Syria,* for example—a game in which users become rebels in the Syrian uprising—was rejected by the App store because it supposedly targeted part of the Syrian population.

55. Jörg Piringer, *abcdefghijklmnopqrstuvwxyz,* iOS app, 2010, http://joerg.piringer.net/index.php?href=abcdefg/abcdefg.xml&mtitle=projects.

56. Jason Edward Lewis, *What They Speak,* iOS app, September 13, 2012, http://www.poemm.net/projects/speak.html.

57. Jason Edward Lewis, *Migration,* iOS app, October 3, 2012, http://www.poemm.net/projects/migration.html.

58. Jason Edward Lewis, *Smooth Second Bastard,* iOS app, November 19, 2012, http://www.poemm.net/projects/bastard.html.

59. Ibid.

60. Erik Loyer, *Strange Rain,* iOS app, updated January 7, 2013, http://opertoon.com/2010/11/strange-rain-for-ipad-iphone-ipod-touch.

61. Ibid.

62. Ibid.

63. McKenzie Wark, *A Hacker Manifesto* (Cambridge, Mass.: Harvard University Press, 2004), 004, 006.

64. Certainly, even the typewriter-based dirty concrete poems and typestracts I discuss in chapter 3 are also codeworks of a sort whose aesthetics are perfectly aligned with the DIY, open-source hacker ethos of certain digital literature practitioners. Rita Raley rightfully points out in her essay "Interferences: [Net.Writing] and the Practice of Codework" that the practice of codework has a long artistic and literary history that predates the Web and even the personal computer, "including Oulipo's Algol code poems and the use of computer instructions in their texts; the long-term tradition of generative aesthetics and poetic programming, such as Tristan Tzara's 'algorithm' for Dadaist composition (including in a similar vein La Monte Young's and John Cage's instructional scores); ASCII art; the composition of Quines; and Perl poems." The difference between a typestract by, say, Dom Sylvester Houédard and a codework by Nick Montfort is, however, that such a gesture of openness in the current era of computing is intensely and explicitly political, oppositional, and activist against corporate control of creative expression, which did not exist in its current form prior to the 1980s. Rita Raley, "Interferences: [Net.Writing] and the Practice of Codework," *electronic book review,* September 8, 2002, http://www.electronicbookreview.com/thread/electropoetics/net.writing.

65. Deena Larsen, *Samplers: Nine Vicious Little Hypertexts,* CD-ROM and user's guide (Watertown, Mass.: Eastgate Systems, 1997).

66. Deena Larsen, e-mail message to author, January 12, 2013.

67. Larsen, *Samplers,* user's guide, 15.

68. Mark Bernstein, "Patterns of Hypertext," Eastgate Systems website, 1999, http://www.eastgate.com/patterns/Patterns.html.

69. Mark Bernstein, "Jane's Spaces," December 28, 2001, http://www.markbernstein.org/December01/Janesspaces.html.

70. Victor Shklovsky, "Art as Technique," *Russian Formalist Criticism: Four Essays,* 2nd ed., trans. Lee T. Lemon and Marion J. Reis (Lincoln: University of Nebraska Press), 12.

71. Olga Goriunova and Alexei Shulgin, "Glitch," in *Software Studies: A Lexicon,* ed. Matthew Fuller (Cambridge, Mass.: MIT Press, 2008), 111.

72. Its relation to an aesthetics of failure and to the embrace of chance means that glitch aesthetics clearly finds its roots in early twentieth-century avant-garde experiments in art, writing, theater, and music. These experiments, on the one hand, sought to disrupt the status quo, which

was supposedly maintained by tranquil, harmonious art, and, on the other hand, reflected a search for a new realism—one that represented the noise and the chaos of a rapidly industrializing world. Luigi Russolo, for example, wrote the Futurist manifesto "The Art of Noises" in 1913, in which he declares: "Today music, as it becomes continually more complicated, strives to amalgamate the most dissonant, strange and harsh sounds. . . . This musical evolution is paralleled by the multiplication of machines, which collaborate with man on every front. Not only in the roaring atmosphere of major cities, but in the country too, which until yesterday was totally silent, the machine today has created such a variety and rivalry of noises that pure sound . . . no longer arouses any feeling." Russolo believed, then, that noise—random, dissonant, machine-based sounds as opposed to what he called "pure sound"—was fast becoming the only way to experience the world anew. Luigi Russolo, "The Art of Noises," unknown.nu, http://www.unknown.nu/futurism/noises.html.

73. Goriunova and Shulgin, "Glitch," 114.

74. Tilman Baumgaertel, "Interview with Jodi," nettime, August 28, 1997, http://www.nettime.org/Lists-Archives/nettime-l-9708/msg00112.html.

75. Jodi, *Untitled Game,* http://www.untitled-game.org.

76. A more recent, lesser-known example of glitch art as an act of hacking the closed, slick interface is Benjamin Gaulon's DIY online data corruption software called *Kindleglitcher,* which allows users to load and glitch e-books. While the EPUB format is a free and open e-book standard, Amazon's Kindle is most certainly not open to users to tinker or create with. Benjamin Gaulon, "RECYCLISM :: KINDLEGLITCHER," http://recyclism.com/kindleglitcher.php.

77. William Gibson and Dennis Ashbaugh, *Agrippa (A Book of the Dead)* (New York: Kevin Begos, 1992).

78. Matthew Kirschenbaum, *Mechanisms: New Media and the Forensic Imagination* (Cambridge, Mass.: MIT Press, 2008), 231.

79. Ibid., 221.

80. William Gibson, *Agrippa (A Book of the Dead),* Source Code website, http://www.williamgibsonbooks.com/SOURCE/AGRIPPA.ASP.

81. Talan Memmott, "Lexia to Perplexia," *Electronic Literature Collection* 1 (October 2006), http://collection.eliterature.org/1/works/memmott__lexia_to_perplexia.html.

82. N. Katherine Hayles, *Writing Machines* (Cambridge, Mass.: MIT University Press, 2002), 49.

83. Memmott, "Lexia to Perplexia."

84. Judd Morrissey, "The Jew's Daughter," *Electronic Literature Collection* 1 (October 2006), http://collection.eliterature.org/1/works/morrissey__the_jews_daughter.html.

85. Jason Nelson, "Game, Game, Game and Again Game," *Electronic*

Literature Collection 2 (February 2011), http://collection.eliterature.org/2/works/nelson_game.html.

86. Jason Nelson, "An Interview with Jason Nelson," *Cordite Poetry Review*, December 1, 2013, http://cordite.org.au/interviews/jason-nelson.

87. Young-Hae Chang Heavy Industries, *Traveling to Utopia: With a Brief History of the Technology*, http://www.yhchang.com/TRAVELING_TO_UTOPIA.html.

88. Hyun-Joo Yoo, "Intercultural Medium Literature Digital: Interview with Young-Hae Chang Heavy Industries," *dichtung-digital* 2 (2005), http://www.dichtung-digital.de/2005/2/Yoo/index-engl.htm.

89. Young-Hae Chang Heavy Industries, "Artist Statement for 'On the Web' (2001–2002)," Anthony Huberman, Curator, PS1 Contemporary Art Center, accessed February 16, 2008, http://www.ps1.org/cut/animations/web/chang.html (webpage discontinued).

90. Young-Hae Chang Heavy Industries, "The Goal Is to Transform a Given Zone: Interview," *avant / garde / under / net / conditions*, http://avantgarde.netzliteratur.net/index.php?bereich=code&aid=149&interview=true.

2. From the Philosophy of the Open to the Ideology of the User-Friendly

1. Related to this shift from the homebrew kit to the user-friendly, GUI-based personal computer is the initial attempt to make computers appear friendly to uncertain first-time buyers by marketing them as sophisticated typewriters. For example, Don Lancaster declares in his *TV Typewriter Cookbook* that his 1973 TV Typewriter can "convert an ordinary Selectric *office* typewriter into a superb hard-copy printer." And a 1979 advertisement in *Byte* magazine for the word processor AUTOTYPE—"a true processor of words"—oddly includes images of text in the shape of arrows and trees that could be easily mistaken for the typewriter-created concrete poetry I discuss in chapter 3. Don Lancaster, *The TV Typewriter Cookbook* (Indianapolis: H. W. Sams, 1976), 218; "AUTOTYPE," advertisement, *Byte*, November 1979.

2. Jean-Louis Gassée, *The Third Apple: Personal Computers and the Cultural Revolution* (New York: Harcourt Brace Jovanovich Publishers, 1985), 11.

3. In 1985, for example, John Whiteside et al. wrote, "Interface style is not related to performance or preference (but careful design is)," and further, they concluded, "The care with which an interface is crafted is more important than the style of interface chosen, at least for menu, command, and iconic systems." John Whiteside, Sandra Jones, Paul S. Levy, and Dennis Wixon, "User Performance with Command, Menu, and Iconic Interfaces," *CHI 1985 Proceedings*, April 1985, 185, 190.

4. Wendy Chun, *Programmed Visions: Software and Memory* (Cambridge, Mass.: MIT Press, 2011), 8.

5. From this perspective, it is, then, no coincidence that Apple had actually designed something like an iPhone in 1983. At the same time that Macintosh designers were hard at work, Hartmut Esslinger, the designer of the Apple IIc, built a white landline phone complete with a built-in, stylus-driven touch screen. The Apple IIc was in fact a close relative of the Macintosh in terms of portability and lack of internal expansion slots, which made them both closed systems. The IIc was also released in 1984, just three months after the Macintosh. Zachary Sniderman, "Apple's First iPhone Was Made in 1983," *Mashable*, December 30, 2011, http://mashable.com/2011/12/30/apple-iphone-1983.

6. McKenzie Wark, *A Hacker Manifesto* (Cambridge, Mass.: Harvard University Press, 2004), 004, 006.

7. Steven Johnson, *Interface Culture: How New Technology Transforms the Way We Create and Communicate* (New York: Basic Books, 1997), 14.

8. Matthew Fuller, *Behind the Blip: Essays on the Culture of Software* (Brooklyn, N.Y.: Autonomedia, 2003), 149.

9. Douglas Engelbart and William English, "A Research Center for Augmenting Human Intellect," in *The New Media Reader*, ed. Noah Wardrip-Fruin and Nick Montfort (Cambridge, Mass.: MIT Press, 2003), 233.

10. Douglas Engelbart, "Augmenting Human Intellect: A Conceptual Framework," in *The New Media Reader*, ed. Noah Wardrip-Fruin and Nick Montfort (Cambridge, Mass.: MIT Press, 2003), 95.

11. Noah Wardrip-Fruin, introduction to "A Research Center for Augmenting Human Intellect," by Douglas Engelbart, in *The New Media Reader*, ed. Noah Wardrip-Fruin and Nick Montfort (Cambridge, Mass.: MIT Press, 2003), 232.

12. Engelbart and English, "A Research Center," 236–37. It's worth pointing out that neither view control nor chains of view are possible on today's word processors or commercially available computers.

13. Steven Johnson rightly points out that Engelbart did introduce "another [graphical] layer separating the user from his or her information." Keeping in mind that not all GUIs are equal, however, Engelbart's interface did not just make it seem "as though the information was now closer at hand" but—*in spite of* this additional layer separating the user from the underlying flow of information—gave the user a greater ability to access that information and, most important, to use that information for tool building than was possible with the command-line interface or, as I explain later in this chapter, the Macintosh GUI. In other words, an additional graphical layer does not necessarily mean less access to information and tool building. Johnson, *Interface Culture*, 21.

14. "What is Logo?," Logo Foundation, 2011, http://el.media.mit
.edu/logo-foundation/logo/index.html.

15. Seymour Papert, *Mindstorms: Children, Computers, and Powerful
Ideas* (New York: Basic Books, 1980), 19.

16. "LOGO," advertisement, *Byte*, February 1982, 255. I focus on *Byte*
throughout this chapter, as it was one of the most influential, technical
computing magazines of the era.

17. Papert, *Mindstorms*, 101.

18. Alan Kay and Adele Goldberg, "Personal Dynamic Media," in *The
New Media Reader*, ed. Noah Wardrip-Fruin and Nick Montfort (Cambridge, Mass.: MIT Press, 2003), 393; emphasis mine.

19. Alan Kay, "User Interface: A Personal View," in *The Art of Human–
Computer Interface Design*, ed. Brenda Laurel (Reading, Mass.: Addison-
Wesley Publishing,1990), 195. Kay documents Papert's influence on him,
writing, "The work of Papert convinced me that whatever user interface
design might be, it was solidly intertwined with learning."

20. Mike Wilber and David Fylstra, "Homebrewery vs the Software
Priesthood," *Byte*, October 1976, 90.

21. Ibid., 90.

22. Kay and Goldberg, "Personal Dynamic Media," 394; emphasis
mine. This belief in the existence of two warring factions in computing
and the necessity to overthrow the so-called software priesthood with
educated, empowered users is echoed again in the same year, 1977, in
Theodor Nelson's *The Home Computer Revolution* (a lesser-known but
more user-friendly version of the 1974 *Computer Lib/Dream Machines*).
Writing vehemently, if not militantly, against "the first computer age
[that] brought us the oppression and bureaucratic abrasion of unpleasant requirements and delays, bad forms and categories, incomprehensible directions, electronic excuses," Nelson proclaims, "The second computer age will be the opposite. . . . We can undo the oppressions of the
first: eliminating paperwork, eliminating the bureaucratic functionaries
who specialized in insulting you for not understanding the incomprehensible, and making things available, comprehensible, open. But the
tradition will die hard, though, and so it is up to the public to know what
to demand. . . . For the computer is limited only by your imagination.
Once you understand how wide are the options the computer allows,
you can make up your own computer applications, out of your interests."
Theodor Nelson, *The Home Computer Revolution* (South Bend, Ind.: Nelson, 1977.), 14–17.

23. Kay and Goldberg, "Personal Dynamic Media," 403.

24. Bill Moggridge, *Designing Interactions* (Cambridge, Mass.: MIT
Press, 2007), 159.

25. Kay, "User Interface," 193.

26. Ingalls, "Design Principles behind Smalltalk," *Byte*, special issue, 1981, 286. Ingalls writes, "If a system is to serve the creative spirit, it must be entirely comprehensible to a single individual. . . . Any barrier that exists between the user and some part of the system will eventually be a barrier to creative expression. Any part of the system that cannot be changed or that is not sufficiently general is a likely source of impediment" (286).

27. Alan Kay, *The Early History of Smalltalk*, http://worrydream.com/ EarlyHistoryOfSmalltalk.

28. Ibid., 83.

29. Ibid., 84.

30. Trygve Reenskaug, "User-Oriented Descriptions of Smalltalk Systems," *Byte*, August 1981, 166.

31. Adele Goldberg, "Introducing the Smalltalk-80 System," *Byte*, August 1981, 18.

32. Ibid.

33. Jeff Johnson, Teresa L. Roberts, William Verplank, David C. Smith, Charles H. Irby, Marian Beard, and Kevin Mackey, "The Xerox Star: A Retrospective," *Computer*, September 1989, 22.

34. Douglas Engelbart, "Workstation History and the Augmented Knowledge Workshop," Doug Engelbart Institute, 2008, http://www. dougengelbart.org/pubs/augment-101931.html. In 1983 *Byte* magazine featured the Apple Lisa by including an interview with the designers of the Lisa. When the interviewer asks, "Do you have a Xerox Star here that you work with?" Larry Tesler responds by saying, "No, we didn't have one here. We went to the NCC [National Computer Conference] when the Star was announced and looked at it. And in fact it did have an immediate impact. A few months after looking at it we made some changes to our user interface based on ideas that we got from it. For example, the desktop manager we had before was completely different; it didn't use icons at all, and we never liked it very much. We decided to change ours to the icon base. That was probably the only thing we got from Star, I think. Most of our Xerox inspiration was Smalltalk rather than Star." Chris Morgan, Gregg Williams, and Phil Lemmons, "An Interview with Wayne Rosing, Bruce Daniels, and Larry Tesler: A Behind-the-Scenes Look at the Development of Apple's Lisa," *Byte*, February 1983, 108.

35. David Canfield Smith, Charles Irby, Ralph Kimball, and Bill Verplank, "Designing the Star User Interface," *Byte* April 1982, 248.

36. Ted Linden, Eric Harslem, and Xerox Corporation, *Office Systems Technology: A Look into the World of the Xerox 8000 Series Products: Workstations, Services, Ethernet, and Software Development* (Palo Alto, Calif.: Office Systems Division, 1982), 11.

37. Steve Wozniak, "The Apple-II," *Byte*, May 1977, 40.

38. "How to Buy a Personal Computer," advertisement, *Byte,* March 1977, 5.

39. John Cage, "Diary: Audience 1966," in *Multimedia: From Wagner to Virtual Reality,* ed. Randall Packer and Ken Jordan (New York: Norton, 2001), 92. That said, there are also examples of writers tinkering with a broad range of computing platforms and systems beyond the Apple II in ways that in their experimental nature also exemplify an activist media poetics. One particular example that has yet to receive any significant critical attention is the work done by the literary collective Invisible Seattle, whose members (writers as well as actors and visual artists) thought of themselves as both literary workers and cultural activists and who dressed in hard hats and white overalls and wandered the streets of Seattle soliciting words and texts from passersby. These words and texts were then pieced together into a novel that member Philip Wohlstetter described as "a gigantic literary computer called Scheherazade II." Scheherazade II was in fact an IBM PC clone that starting in 1983 ran an electronic bulletin board called IN.S.OMNIA, to which users could log on and contribute more text to the collaborative, distributed novel of Invisible Seattle. Wittig evocatively describes this literary BBS as "a zone where certain familiar functions of reading and writing seem to occur, but in conditions that are completely new. What kind of writing is this that parades a glowing alphabet across a black screen? What kind of writing is this that is evanescent, that exists in front of you, then ceases to exist at the touch of a key?" Philip Wohlstetter, foreword to *Invisible Rendezvous: Connection and Collaboration in the New Landscape of Electronic Writing,* by Rob Wittig (Middletown, Conn.: Wesleyan University Press, 1994), ix, x; Rob Wittig, *Invisible Rendezvous: Connection and Collaboration in the New Landscape of Electronic Writing* (Middletown, Conn.: Wesleyan University Press, 1994), 3, 17.

40. Jeremy Reimer, "Total Share: 30 Years of Personal Computer Market Share Figures," *Ars Technica* blog, December 14, 2005, http://arstechnica.com/old/content/2005/12/total-share.ars/1.

41. Nichol distributed 100 signed copies on 5.25-inch floppies through Underwhich, a publishing collective he helped to start in 1979, along with Steve McCaffery, Paul Dutton, Steven Ross Smith, John Riddell, and Richard Truhlar. Once the Apple IIe became obsolete in the early 1990s, *First Screening* was republished as a HyperCard version on 3.5-inch floppies in 1993 by Red Deer College Press and has been recently made available as an emulated version thanks to the efforts of Jim Andrews, Geof Huth, Lionel Kearns, and Dan Waber. bpNichol, *First Screening,* Vispo, http://vispo.com/bp.

42. "Printed Matter from *First Screening,*" Vispo, http://vispo.com/bp/emulatedversion.htm#pm.

43. Jim Andrews, "Framing 'Off-Screen Romance,'" Vispo, http://vispo.com/bp/jim.htm.

44. Geof Huth, "Endemic Battle Collage," Vispo, http://vispo.com/huth/endemic_battle_collage.htm; Geof Huth, e-mail message to author, March 24, 2012.

45. Paul Zelevansky, *The Case for the Burial of Ancestors: Book Two: Genealogy* (New York : Zartscop, 1986), ix.

46. Ibid.

47. Ibid., xii.

48. Forth was a popular programming language for home computers that had particularly limited memory.

49. Zelevansky, *Book Two*, ix.

50. Ibid., 2.

51. Ibid., 2.

52. Ibid., 9.

53. Ibid., 9.

54. Andy Hertzfeld, *Revolution in the Valley* (Sebastopol, Calif.: O'Reilly, 2005), xviii; Steven Levy, *Insanely Great: The Life and Times of Macintosh, the Computer That Changed Everything* (New York: Viking, 1994), 8, 157.

55. Chris Rutkowski, "An Introduction to the Human Applications Standard Computer Interface: Part 1: Theory and Principles," *Byte*, October 1982, 291, 299–300.

56. Sherry Turkle alternatively frames this transformation in the use of the word *transparency* in terms of our culture's turn toward simulation such that by the end of the 1980s, *transparency* had "taken on its Macintosh meaning in both computer talk and colloquial language. In a culture of simulation, when people say that something is transparent, they mean that they can easily see how to make it work. They don't necessarily mean that they know why it is working in terms of any underlying process." Sherry Turkle, *Life on the Screen: Identity in the Age of the Internet* (New York: Simon & Schuster, 1995), 42.

57. Rutkowski, "An Introduction," 390.

58. Levy, *Insanely Great*, 228.

59. Larry Tesler, "The Legacy of the Lisa," *Macworld*, September 1985, 17.

60. Hertzfeld, *Revolution*, 50.

61. Levy, *Insanely Great*, 228.

62. "'1984' Apple Macintosh Commercial," YouTube video, 0:59, from an advertisement originally aired on CBS on January 22, 1984, posted by Giancarlo Romeo, October 7, 2008, http://www.youtube.com/watch?v=Z8jioB98IMo.

63. Ibid.

64. That said, Ted Friedman offers an incomparable and far more nuanced reading of Apple's "1984" advertisement in chapter 5 of his *Electric Dreams*. He not only touches on the irony of how the ad "allowed Apple

to harness the visual fascination of a high-tech future while dissociat-
ing itself from its dystopic underside" but also brilliantly teases out the
significance of the fact that the hammer thrower in the ad is a woman,
thereby gendering "the archetypal Mac user as female. And it genders
the Mac *itself* as female." Ted Friedman, *Electric Dreams: Computers in
American Culture* (New York: NYU Press, 2005), 112.

65. Quoted in Levy, *Insanely Great*, 164–65.

66. Apple Computer, *Apple Human Interface Guidelines: The Apple
Desktop Interface* (Reading, Mass.: Addison-Wesley Publishing, 1987), 2.

67. Ibid., 14.

68. Remarkably, Apple designer Thomas D. Erickson later described
the Macintosh interface in somewhat negative terms, as he called it a
kind of "pidgin" insofar as it had a very simple noun-verb syntax: "First
you select the object, then you specify the action to be carried out. The
Macintosh interface has almost no tense," and so it also "has distinct
limitations in its communicative power—you can get your basic tasks
done, but that's about it." Yet as I explain in chapter 1, this limited interface originally targeted at "users who lacked the time or inclination
to learn about a computer" has become the de facto computing interface
for all users. Thomas D. Erickson, "Interface and the Evolution of Pid-
gins: Creative Design for the Analytically Inclined," in *The Art of Human–
Computer Interface Design*, ed. Brenda Laurel (Reading, Mass.: Addison-
Wesley Publishing, 1990), 13.

69. Levy, *Insanely Great*, 4.

70. Ibid., 5.

3. Typewriter Concrete Poetry as Activist Media Poetics

1. Siegfried Zielinski, *Deep Time of the Media: Toward an Archaeology
of Hearing and Seeing by Technical Means* (Cambridge, Mass.: MIT Press,
2006), 3.

2. Ibid., 7, 10.

3. Peter Finch, *Typewriter Poems* (Millerton, N.Y.: Something Else
Press, 1972), 47.

4. Extended critical readings of the typewriter as a reading/writing
medium have so far been few and far between. Other than Darren
Wershler's 2007 *The Iron Whim: A Fragmented History of Typewriting*, the
English-language version of Kittler's 1999 *Gramophone, Film, Typewriter,*
which follows his more diffuse writing on the typewriter throughout *Discourse Networks, 1800/1900*, does have a substantial chapter dedicated to
so-called old media such as the typewriter, which he frames with the
notorious claim that "understanding media—despite McLuhan's title—
remains an impossibility precisely because the dominant information
technologies of the day control all understanding and its illusions."

While he does take from McLuhan the conviction that media unavoidably transform us, citing Nietzsche as an example of one who "changed from arguments to aphorisms, from thoughts to puns, from rhetoric to telegram style" once he started to use a typewriter, his account of the typewriter, on the one hand, is dominated by his conviction that in the "convergence of a profession, a machine, and a sex," it "only inverts the gender of writing" and, on the other hand, despite his passing observations that "poésie concrète" is a "form of pure concrete poetry" and that "Remington's and Underwood's invention ushered in a poetics that William Blake or John Donne with their limits/ears could not hear," is simply not concerned with the ramifications of this writing machine on the broader twentieth-century literary landscape. Darren Wershler-Henry, *The Iron Whim: A Fragmented History of Typewriting* (Ithaca, N.Y.: Cornell University Press, 2007); Friedrich Kittler, *Gramophone, Film, Typewriter*, trans. Geoffrey Winthrop-Young (Stanford, Calif.: Stanford University Press, 1999); Friedrich Kittler, *Discourse Networks, 1800/1900*, trans. Michael Metteer and Chris Cullens (Stanford, Calif.: Stanford University Press, 1990), xxxix, 203, 183, 229, 231.

5. Marshall McLuhan, *Understanding Media: The Extensions of Man* (Cambridge, Mass.: MIT Press, 1994).

6. Peggy Curran, "McLuhan's Legacy Is Alive and Tweeting," *Montreal Gazette*, July 17, 2011, http://www2.canada.com/reginaleaderpost/news/weekender/story.html?id=68d18c46-d3e8-4b06-b737-4d570ed9cae0. Credit is due entirely to Darren Wershler for introducing me to McLuhan as one engaged in the practice of poetics. "McLuhan," he declares, "writes like a poet, and if you read it like a poem, it makes sense."

7. It is worth remarking that not only was Michel Foucault—whose *The Archaeology of Knowledge* (1969) is commonly touted as a founding text in the field of media archaeology—writing the foregoing at roughly the same time as McLuhan was writing *Verbi-Voco-Visual* but also it appears that the roots of his archaeological thought lie in his 1966 *The Order of Things* and extend beyond *The Archaeology of Knowledge* to his 1973 *This Is Not a Pipe*. In the context of this chapter, what is remarkable is that Foucault's description of heterotopias in *The Order of Things* seems to precisely describe all that dirty concrete undoes—from its turn away from semantic meaning to its use of the typewriter as a means to turn words into objects one ought to look at rather than read. He writes that unlike utopias, which "afford consolation," *heterotopias* are "disturbing . . . probably because they secretly undermine language, because they make it impossible to name this *and* that, because they shatter or tangle common names, because they destroy 'syntax' in advance. . . . Heterotopias . . . desiccate speech, stop words in their tracks, contest the very possibility of grammar at its source; they dissolve our myths and sterilize the lyricism of our sentences." I touch on *The Archaeology of*

Knowledge in chapter 4, but here I want to point out that *This Is Not a Pipe* not only seems to carry on his archaeological line of thought but does so with an anachronistic reading of visual works by René Magritte. Of particular note for this chapter on concrete poetry is his chapter on the calligramme in Magritte's paintings, which he reminds us, has the ability to both draw attention to the page as a writing interface that's anything but blank or empty of meaning and offer us a McLuhanesque verbi-voco-visual experience: "It distributes writing in a space no longer possessing the neutrality, openness, and inert blankness of paper. It forces the ideogram to arrange itself according to the laws of a simultaneous form. For the blink of an eye, it reduces phoneticism to a mere grey noise completing the contours of the shape; but it renders outline as a thin skin that must be pierced in order to follow, word for word, the outpouring of its internal text. . . . Thus the calligram aspires playfully to efface the oldest oppositions of our alphabetical civilization: to show and to name; to shape and to say; to reproduce and to articulate; to imitate and to signify; to look and to read." These earlier works by Foucault are remarkably similar to the underrecognized works by McLuhan that use poetry and poetics to inflect the study of media. Both seem to suggest that formally experimental poetry and poetics could be reimagined as engaged with media studies, a fundamental shift in perspective that bears with it the potential to make poetics a relevant means by which to understand (Western, English-language) digital textuality. Michel Foucault, *The Archaeology of Knowledge*, trans. A. M. Sheridan Smith (New York: Random House, 1972); Michel Foucault, *The Order of Things: An Archaeology of the Human Sciences* (New York: Random House, 1994), xviii; Michel Foucault, *This Is Not a Pipe*, trans. and ed. James Harkness (Berkeley: University of California Press, 1982), 21.

8. The first collection of concrete poetry appeared in 1957 not as an anthology but as the inaugural issue of the journal *Material,* published by the German poet Daniel Spoerri. This issue featured contributions by Josef Albers, Louis Aragon, Helmut Heissenbüttel, Eugen Gomringer, Dieter Roth, and others.

9. Dick Higgins, "Statement on Intermedia" Artpool, http://www.artpool.hu/Fluxus/Higgins/intermedia2.html.

10. Augusto de Campos, Decio Pignatari, and Haroldo de Campos, "Pilot Plan for Concrete Poetry," in *Concrete Poetry: A World View,* ed. Mary Ellen Solt (Bloomington: Indiana University Press, 1970), 71.

11. McLuhan was surely aware of Robert Creeley's declaration that "form is never more than extension of content" as it was popularized in Charles Olson's 1950 "Projective Verse."

12. Marshall McLuhan, V. J. Papanek, J. B. Bessinger, Karl Polanyi, Carol C. Hollis, David Hogg, and Jack Jones. *Verbi-Voco-Visual Explorations* (New York: Something Else Press, 1967), unpaginated.

13. Ibid., unpaginated.

14. Ibid., 16.

15. Ibid., 16–17; emphasis my own.

16. Marshall McLuhan, *Culture Is Our Business* (New York: McGraw-Hill, 1970), 44.

17. Zielinski, *Deep Time*, 256.

18. Wershler-Henry, *The Iron Whim*, 242.

19. Charles Olson, "Projective Verse," in *Collected Prose*, ed. Donald Allen and Benjamin Friedlander (Berkeley, Calif.: University of California Press, 1997), 245.

20. Ibid.

21. Steve McCaffery is, in fact, one of the only concrete poets who openly works against this conventional reading of Olson's use of the typewriter for a breath-based poetics. In fact, in a 1998 interview with Peter Jaeger, McCaffery declares that his dirty concrete poem *Carnival* represents the "repudiation of a breath-based poetics." His explanation of what he was instead attempting seems only to support my reading of Olson as endorsing an emphasis on (the material, mechanical aspect of the) process and the product in poetry whereby both are, in and of themselves, meaningful. McCaffery states that he wanted to extend "the typewriter beyond Olson's own estimation of its abilities (to provide a precise notation of breathing) into a more 'expressionistic' as well as cartographic instrument, approaching the typewriter less as a notational device than a form of saxophone." In other words, he sought to use the typewriter in *Carnival* not as a neutral means of transcription but rather as a means to explore the typewriter's material contours. Steve McCaffery, *Seven Pages Missing: Volume One: Selected Texts 1969–1999* (Toronto: Coach House Books, 2000), 447.

22. Aram Saroyan, untitled, in *An Anthology of Concrete Poetry*, ed. Emmett Williams (New York: Something Else Press, 1967), unpaginated.

23. Dom Sylvester Houédard, untitled, in *An Anthology of Concrete Poetry*, ed. Emmett Williams (New York: Something Else Press, 1967), unpaginated.

24. As I argue in chapter 1, despite its disinterest in semantic meaning, such poetry does not lack meaning altogether—rather, it is invested in broadening our sense of meaning to include a thoroughgoing exploration of materiality as meaning and, as such, is also a clear precursor to contemporary digital texts such as those by Judd Morrissey and Jason Nelson.

25. Dom Sylvester Houédard, untitled, unpaginated.

26. Mary Ellen Solt, ed., *Concrete Poetry: A World View* (Bloomington: Indiana University Press, 1970), 34.

27. Ronald Johnson as quoted in Solt, *Concrete Poetry*, 52; emphasis my own.

28. Ronald Johnson, "MAZE," in *Concrete Poetry: A World View,* ed. Mary Ellen Solt (Bloomington: Indiana University Press, 1970), 251.

29. Dick Higgins as quoted in Solt, *Concrete Poetry,* 57.

30. Marjorie Perloff, *Radical Artifice: Writing Poetry in the Age of Media* (Chicago: University of Chicago Press, 1991), 232; Caroline Bayard as quoted in Perloff, *Radical Artifice,* 232.

31. Olson, "Projective Verse," 240.

32. Perloff, *Radical Artifice,* 114. Perloff only points out that Steve McCaffery used the term to describe his *Carnival* and that it describes the "later more iconoclastic version" of concrete poetry.

33. Henri Chopin, untitled, *Stereo Headphones: an occasional magazine of the new poetries* 1 (Spring 1970): unpaginated. It is worth noting that in the same issue of *Stereo Headphones,* Pierre Garnier counters Chopin by writing, "I definitely do not believe that spatial and concrete poetry is dead, but I think that we must get our breath back and above all assimilate (before using it in our poetry) everything that has happened to the world during the past year. New problems have appeared—especially at a political level—towards which poets can not remain indifferent."

34. As Steve McCaffery wrote to the Poetics Listserv in late February 2011: "'Dirty' concrete as I recall is a term like 'Dada' with an uncertain origin. It was a familiar usage in the early seventies in my own discussions with bp Nichol about the incipient hierarchization within the international concrete movement. We both noted that anthologies were regurgitating the same material which was straight edged, typographically lucid (Garnier's work for instance and Eugen Gomringer's as well as Ian Hamilton Finlay's in Scotland and that of the de Campos brothers in Brazil). We both considered that what seemed to offer itself as a vanguard movement dedicated to poetic change was rapidly ossifying. Nichol certainly used the term in his letters to and from Stephen Scobie (living in the same town as bp we met and chatted rather than wrote to each other hence my take is purely anecdotal)." Steve McCaffery to Poetics Listserv, February 28, 2011.

35. "Dirty concrete" was first used either by bpNichol, bill bissett, or Stephen Scobie; the term is almost certainly Canadian in origin. There are, however, no documents that prove this definitively. I am tremendously grateful to George Bowering, Jack David, Frank Davey, Jamie Hilder, Steve McCaffery, Stephen Scobie, and Darren Wershler for their attempts to help me track down the history of the term.

36. Stephen Scobie to bpNichol, June 26, 1968, Special Collections, Simon Fraser University Libraries.

37. Writes Jean-Francois Bory in this issue of *Stereo Headphones*: "I think one can define concrete poets as those appearing in the big concrete poetry anthologies. . . . I am very much afraid that we've witnessed an ignoble recuperative operation in these two anthologies, into which

everybody threw themselves because it suited them to do so at the time. . . . So? . . . So the word concrete, which clearly represents no concept at all, has through the pressure of various authors, become part of history." Jean-François Bory, untitled, *Stereo Headphones: an occasional magazine of the new poetries* 1 (Spring 1970).

38. bpNichol, "concrete," in *Meanwhile: The Critical Writings of bpNichol,* ed. Roy Miki. (Vancouver, BC: Talonbooks, 2002), 30.

39. Jack David, "Visual Poetry in Canada: Birney, bissett, and bp," *Studies in Canadian Literature* 2, no. 2 (1977), http://journals.hil.unb.ca/index.php/SCL/article/view/7870/8927. I should also note that "dirty concrete" was later picked up in Stephen Scobie's 1984 book-length study *bpNichol: What History Teaches,* in which he aligns Mike Weaver's use of the terms "expressionist" and "constructivist" with Ian Hamilton Finlay's "suprematist" and "fauvre" and claims that "more simply, bpNichol spoke of a division between 'clean' and 'dirty' concrete." When I wrote to Stephen Scobie asking whom he thought was the originator of "dirty concrete," he responded, however, that he had heard it from the English critic Mike Weaver, who was writing very early on about concrete poetry. I conducted a phone interview with Mike Weaver in March 2011 in which he denied using the term. Stephen Scobie, *bpNichol: What History Teaches*, (Vancouver, BC: Talonbooks, 1984), 35, 139.

40. David, "Visual Poetry in Canada." In the same article, David claims that Rosalie Murphy refers to "dirty concrete" in her 1970 *Contemporary Poets of the English Language*. I inspected the Murphy book, however, and could not find any reference to either dirty or clean concrete poetry. In fact, Frank Davey informed me in an e-mail correspondence that David is in fact referring to Davey's own 1971 definition of clean and dirty concrete he includes in *Earle Birney* and that he wrote with the assistance of bpNichol: "Concrete is usually divided by its devotees into 'clean' and 'dirty'. In clean concrete, the preferred and dominant type, the visual shape of the work is primary, the linguistic signs secondary. In this view the most effective concrete poems are those with an immediate and arresting visual effect which is made more profound by the linguistic elements used in the poem's constituent parts. The weakest are dirty concrete, those with amorphous visual shape and complex and involute arrangements of linguistic elements. In dirty concrete there can be no immediate to the whole, only a cumulative interpretation gained by painstaking labour." In the same e-mail correspondence, Frank Davey writes, "When I met bp in 1970, he told me that clean concrete was a kind you could understand by looking but not reading, and that dirty was the kind that had a visual shape made of phrases or clauses or sentences that had to be read as well as viewed. (But he didn't attribute that theory to anyone.)" Rosalie Murphy, ed., *Contemporary Poets of the English Language* (Chicago: St. James Press, 1970), 99; Frank Davey, *Earle Birney*

(Toronto: Copp Clark Publishing, 1971), 65; Frank Davey, e-mail message to author, March 1, 2011.

41. bill bissett, "a pome in praise of all quebec bombers," in *pass th food release th spirit book* (Vancouver, BC: Talonbooks, 1973), unpaginated.

42. bill bissett, e-mail message to author, April 21, 2011.

43. Steve McCaffery, "McLuhan + Language × Music," in *North of Intention: Critical Writings 1973–1986* (New York: Roof Books, 2000), 85.

44. bpNichol, "statement, november 1966," in *Meanwhile: The Critical Writings of bpNichol*, ed. Roy Miki (Vancouver, BC: Talonbooks, 2002), 18.

45. Marshall McLuhan, "The Medium Is the Message," in *Understanding Media: The Extensions of Man* (Cambridge, Mass.: MIT Press, 1994), 8. Nichol similarly framed this desire to move beyond an ego-based poetry several years later, in a 1973 letter to Mary Ann Solt, as one that came out of his belief that up to that point he was "too arrogant": "[i] had found myself trying to dominate in the act of writing the language i was using as opposed to letting myself simply learn from the signs themselves." bpNichol, "A Letter to Mary Ellen Solt," in *Meanwhile: The Critical Writings of bpNichol*, ed. Roy Miki (Vancouver, BC: Talonbooks, 2002), 116.

46. Steve McCaffery and bpNichol, *Rational Geomancy: The Kids of the Book-Machine* (Vancouver, BC: Talonbooks, 1992), 141; emphasis my own.

47. Concrete poets also exploited the printing and distribution capabilities of the mimeograph machine as a crucial extension of their politically oriented work with the typewriter. In an essay on the close ties between Nichol and the Cleveland-based dirty concrete mimeoist d.a.levy, Douglas Manson rightly points out, "The mimeo revolution in poetry was also the typewriter revolution, because it had supplanted the compositor's box and the laying of type for a press, and instead utilized mimeograph stencils to transfer the typed manuscript into the appearance of the page in the published work." More, since mimeograph stencils were notorious for quickly deteriorating, the degraded quality of the image made the mimeograph machine an ideal printing and distribution extension of dirty concrete poetry created with a typewriter. Douglas Manson, "Mimeograph as the Furnace of Loss: The Literary Friendship of bpNichol and d.a.levy," *d.a.levy & the mimeograph revolution*, ed. Larry Smith and Ingrid Swanberg (Huron, Ohio: Bottom Dog Press, 2007), 194.

48. Marshall McLuhan, "Culture and Technology," in *Astronauts of Inner-Space: A Collection of Avant-Garde Activity*, ed. Jeff Berner (San Francisco, Calif.: Stolen Paper Review, 1966), 18. This same collection provides textual evidence that even concrete poets from the first generation such as Decio Pignatari (who was both a poet and a professor of information theory) were attempting to make their work more politically engaged by way of a thoroughgoing experimentation with form that, in the case of concrete poetry, was inextricable from writing media or from its material means of production. Pignatari wrote in 1966, the same year

NOTES TO CHAPTER 3

as Nichol's "Statement": "From 1961 on, concrete poets face definitely the 'engagement' question. What issued—social and political concrete poetry—was chiefly based on Mayakofsky: 'There is no revolutionary art without revolutionary form.'" Decio Pignatari, "The concrete Poets of Brazil," in *Astronauts of Inner-Space*, ed. Jeff Berner, 8.

49. bpNichol, "Interview: Raoul Duguay," in *Meanwhile: The Critical Writings of bpNichol*, ed. Roy Miki (Vancouver, BC: Talonbooks, 2002), 120.

50. Nelson Ball, introduction to *Konfessions of an Elizabethan Fan Dancer*, by bpNichol (Toronto, ON: Coach House Books, 2004), 10.

51. bpNichol, "The Complete Works," pamphlet (Toronto, ON: Ganglia, 1968).

52. Aram Saroyam, "The Collected Works," in *Complete Minimal Poems* (New York: Ugly Duckling Press, 2007), 151.

53. bpNichol, "Interview: Caroline Bayard and Jack David," in *Meanwhile: The Critical Writings of bpNichol*, ed. Roy Miki (Vancouver, BC: Talonbooks, 2002), 171; emphasis my own.

54. bpNichol, *Translating Translating Apollinaire: A Preliminary Report from a Book of Research* (Milwaukee, Wis.: Membrane Press, 1979), unpaginated.

55. bpNichol, *Sharp Facts: Some Selections from Translating Translating Apollinaire 26* (Milwaukee, Wis.: Membrane Press, 1980), unpaginated.

56. Ibid., unpaginated, emphasis my own.

57. bpNichol, "The Medium Was the Message," in *Meanwhile: The Critical Writings of bpNichol*, ed. Roy Miki (Vancouver, BC: Talonbooks, 2002), 300.

58. Most of the Four Horsemen's creative output simultaneously explored the verbi-, the voco-, and the visual through nonrepresentational and asyntactic sounds, improvisation, and a loose method for visually notating performance scores for these sounds. Thus, it is noteworthy—though not surprising—that according to Steve McCaffery, Marshall McLuhan owned all of the Four Horsemen LPs. While no doubt McLuhan's interest in the Four Horsemen partly stemmed from his fascination with the "secondary orality" he believed defined the electric age, he was also in personal contact with Four Horsemen member Paul Dutton, as McLuhan was a lay participant in the liturgy for the church to which both belonged.

59. bpNichol, "Interview: Geoff Hancock," in *Meanwhile: The Critical Writings of bpNichol*, ed. Roy Miki (Vancouver, BC: Talonbooks, 2002), 405.

60. Notably, McCaffery and Nichol include two books by McLuhan in the bibliography for *Rational Geomancy* and *The Interior Landscape: The Literary Criticism of Marshall McLuhan*, both published in 1969. Surprising for McLuhan but presaging Nichol's slightly later conclusions about

typewriting versus handwriting, in *Counterblast* McLuhan writes, "The typewriter is a good distancer. You're less closely attached to what you're writing. Handwriting remains a part of you." Clearly in line with Nichol's and McCaffery's experiments with the way in which the typewriter turns the page into a canvas, McLuhan continues: "It's difficult to see the shape of sentences in the maze of handwriting. When typing, you're more conscious of the appearance of your writing. You view it stretched out before you, detached from you." Marshall McLuhan, *Counterblast* (New York: Harcourt, Brace & World, 1969), 104. McLuhan, ed. Eugene MacNamara, *The Interior Landscape: The Literary Criticism of Marshall McLuhan* (New York: McGraw-Hill, 1969).

61. McCaffery and Nichol, *Rational Geomancy,* 10.

62. Steve McCaffery, *Carnival* (Toronto: Coach House Books, 2001), http://archives.chbooks.com/online_books/carnival/index.html.

63. Steve McCaffery, "Broken Mandala," note in *Seven Pages Missing: Volume Two: Previously Uncollected Texts 1968–2000* (Toronto: Coach House Books, 2002), 439.

64. Steve McCaffery, "Punctuation Poem," in *Seven Pages Missing: Volume One: Selected Texts 1969–1999* (Toronto: Coach House Books, 2000). 56; Steve McCaffery, "Suprematist Alphabet," in *Seven Pages Missing: Volume One,* 65.

65. It is striking that just eight years before McCaffery's "Suprematist Alphabet," Houédard wrote the following about the "wordless suprematist" as an accompaniment to his 1972 typestract "Like Contemplation": "like contemplation, goal of poetry-purge, / was the wordless suprematist / white on white, poeme blanc, / concrete fractures linguistics, atomises words / into incoherence, constricting language / to jewel-like semantic areas where / poet and reader meet / in maximum communication . . . / it is possible to think in images alone— / diagrams, models, gestures and / muscular movements— / as well as in words alone. / poems that are concrete objects themselves, / not windows into souls . . . / concrete poems just ARE; / have no outside reference: / they are objects like TOYS & TOOLS." Dom Sylvester Houédard, "Like Contemplation," UbuWeb Visual Poetry, http://ubumexico.centro.org.mx/text/vp/DSH001_houedard_like_contemplation_1972.pdf.

66. Not surprisingly, Houédard also corresponded with other major writers of the time who were heavily invested in the typewriter—such as Edwin Morgan, Allan Ginsberg, William Burroughs, and Jack Kerouac.

67. Dom Sylvester Houédard, "Leaning on an Angel," Sackner Archive of Concrete and Visual Poetry, http://ww3.rediscov.com/sacknerarchives/ShowItem.aspx?325201355323~34713.

68. "Dom Sylvester Houédard, England, 1924–1992," ubuweb, http://www.ubu.com/historical/houedard/index.html.

69. Dom Sylvester Houédard, "cool poem," Sackner Archive of

Concrete and Visual Poetry, http://ww3.rediscov.com/sacknerarchives/ShowItem.aspx?325201354749~46240.

70. Marshall McLuhan, "Media Hot and Cold," in *Understanding Media: The Extensions of Man* (Cambridge, Mass.: MIT Press, 1994), 23.

71. Dom Sylvester Houédard to bpNichol, July 10, 1965, Special Collections, Simon Fraser University Library.

72. As quoted in Marjorie Perloff, "'Inner Tension / In Attention': Steve McCaffery's Book Art," in *Poetry On and Off the Page: Essays for Emergent Occasions* (Evanston, Ill.: Northwestern University Press, 1998), 267–68.

73. McCaffery, *Carnival.*

74. Steve McCaffery, introduction to *Carnival* (Toronto: Coach House Books, 2001), http://archives.chbooks.com/online_books/carnival/2_introduction.html.

75. That said, it is important to keep in mind that the dirt that defines the analog *Carnival* is simply not present in the digital version. Darren Wershler rightly, and presciently, declared in a 1998 note about his poem "Saint Ratification": "It's extraordinarily difficult to produce a 'dirty' concrete poem on a computer, because the level of pixel-by-pixel manipulation imparts at least the illusion that everything is potentially controllable." The randomness and nonlinearity of *Carnival* communicates anything but control. Darren Wershler, *Nicholodeonline,* January 1998, http://archives.chbooks.com/online_books/nicholodeon/surplus.html.

76. McCaffery, introduction to *Carnival.*

77. Ibid.

78. Christian Bök and Darren Wershler-Henry, "Walls That Are Cracked: A Paralogue on Panels 1 & 2 of Steve McCaffery's *Carnival,*" *Open Letter* 10, no. 6 (1999), 27.

79. It is also the stubbornly lingering evidence of countless alignments, realignments, dealignments of the page, along with the dirt, dust, smudges in the serifs, signs of wear in the keys, misaligned letters, ribbon wear in the inking, and so on that means *Carnival* is a text that *defies* close reading, the very bread and butter of literary studies, in much the same way that many digital poems defy close reading. Rather, the kind of close reading it demands amounts more to description as analysis—as I am attempting to do here—than to a careful, word-by-word analysis. After all, recall McCaffery's injunction: "It's important to remember that the mask excludes and deletes much of the written text." If the written text itself is deliberately occluded, then what we are left with is writing media and the labor of writing itself. Quoted in Perloff, "'Inner Tension / In Attention,'" 267.

80. Casey Reas and Ben Fry, *Processing: A Programming Handbook for Visual Designers and Artists* (Cambridge, Mass.: MIT Press, 2007), 3.

81. John Maeda, "Creative Leaders Get Their Hands Dirty," *Harvard*

Business Review blog, April 20, 2009, http://blogs.hbr.org/maeda/2009/04/the-dirty-mba.html.

82. Reas and Fry, *Processing,* 3.

83. Quoted in Ibid., 3.

84. Mary Flanagan, "[the house]," in *The Electronic Literature Collection,* vol. 1, ed. N. Katherine Hayles, Nick Montfort, Scott Rettberg, and Stephanie Strickland, http://collection.eliterature.org/1/works/flanagan__thehouse.html.

85. Brian Kim Stefans, "The Dreamlife of Letters," in *The Electronic Literature Collection,* vol. 1, ed. N. Katherine Hayles, Nick Montfort, Scott Rettberg, and Stephanie Strickland, http://collection.eliterature .org/1/works/stefans__the_dreamlife_of_letters.html; Brian Kim Stefans, "Letter Builder Update," *Free Space Comix: The Blog,* July 27, 2009, accessed March 24, 2013, http://www.arras.net/fscIII/?p=409 (site discontinued).

86. Quoted in Reas and Fry, *Processing,* 3.

87. Further, this process-oriented ethic of (loosely speaking) open-source making is not limited to dirty concrete poems or certain programming languages—it also underlies a range of recent community-driven artistic/cultural phenomenon such as the demo-scene, the chiptune music scene, Maker Faire, and any of the burgeoning DIY electronics and robotics movements supported by companies such as Makerbot or Arduino. While the impetus for many of these contemporary digital DIY movements can be traced to the arts and crafts movement and the growing prominence of artists books at the end of the nineteenth and the beginning of the twentieth centuries, which were reactions to rapid industrialization and mechanization, the way in which the meaning is in the making as well as in an exploration of surface as depth now seems to be less about the grain of the wood, the binding of the book, the reworking of the physical page and more about how the meaning is in the code, the software, the programming, the circuitry.

88. It was not coincidental that McCaffery in particular was creating *Carnival* during the most lively years of the Homebrew Computer Club or during the most influential years of the DIY-inspired and enormously influential *Whole Earth Catalog*—an ethic of making was in the air. Looking at the *Whole Earth Catalog,* which was published consistently from 1968 to 1972 and then sporadically until 1998, it advocated a philosophy remarkably similar to that embodied by McCaffery's typestracts—a philosophy of making that treated tools as both process and product. As Fred Turner writes, "At one level, the Catalog was a 'Whole Earth' in its own right. That is, it was a seemingly comprehensive informational system, an encyclopedia, a map. . . . At another level, the Catalog offered its readers ways to enter its world and become 'as gods' in a local sense too. The reader could order the 'tools' on display and so help to create a

realm of 'intimate, personal power' in her or his own life. . . . One reader explained the distinction thus: 'Walking to the bathhouse today, holding my new twenty-ounce hammer, I suddenly understood the Whole Earth Catalogue meaning of 'tool.' I always thought tools were objects, things: screw drivers, wrenches, axes, hoes. *Now I realize that tools are a process*: using the right-sized and shaped object in the most effective way to get a job done.'" Fred Turner, *From Counterculture to Cyberculture: Stewart Brand, the Whole Earth Network, and the Rise of Digital Utopianism* (Chicago: University of Chicago Press, 2008), 83; emphasis my own.

89. McCaffery, introduction to *Carnival*.

90. Michael Basinski, "Letter to Larry Smith & Ingrid Swanberg, Concerning d.a.levy's concrete Poetry; Some Ruminating Thoughts Imagined in 2006 (Which Were Some Ruminating Thoughts Once Written to Derek Beaulieu in 2001)," in *d.a.levy & the mimeograph revolution*, ed. Larry Smith and Ingrid Swanberg (Huron, Ohio: Bottom Dog Press, 2007), 236.

4. The Fascicle as Process and Product

1. Wolfgang Ernst might call this particular study of interface an investigation into the "technological conditions of the sayable and thinkable in culture, an excavation of evidence of how techniques direct human or non-human utterances—without reducing techniques to mere apparatuses." Geert Lovink, "Archive Rumblings: Interview with German Media Archaeologist Wolfgang Ernst," Nettime.org, February 25 2003, http://www.nettime.org/Lists-Archives/nettime-l-0302/msg00132.html.

2. Galloway, "The Unworkable Interface," 931.

3. Jeff Han, "Jeff Han: Unveiling the Genius of Multi-touch Interface Design," TED video, 10:11, from a recording made in February 2006, posted by Tedconfjune, August 1, 2006, http://blog.ted.com/2006/08/01/jeff_han_on_ted.

4. "About the NUI Group," NUI Group, http://nuigroup.com/log/about.

5. Not only have I myself produced scholarship that replicates this tendency toward a linear uncovering of origins in digital poetry, but so too have any number of critics. No matter the goals in this book, in media archaeology, or in Foucault's unrealized aspirations for cultural transformation, Western culture understands itself and its past in terms that may very well be impossible to transcend.

6. Michel Foucault, *The Archaeology of Knowledge,* trans. A. M. Sheridan Smith (New York: Random House, 1972), 128.

7. Wolfgang Ernst, "Media Archaeography: Method and Machine versus History and Narrative of Media," in *Media Archaeology,* ed. Jussi

Parikka (Berkeley: University of California Press, 2011), 241; emphasis my own.

8. Drucker echoes this expansive definition of *interface,* writing that "a book is an interface, so is a newspaper page, a bathroom faucet, a car dashboard, an ATM machine." Johanna Drucker, "Humanities Approaches to Interface Theory," *Culture Machine* 12 (2011): 10.

9. Matthew Fuller, *Behind the Blip: Essays on the Culture of Software* (Brooklyn, N.Y.: Autonomedia, 2003), 149.

10. Lovink, "Archive Rumblings."

11. N. Katherine Hayles, *Writing Machines* (Cambridge, Mass.: MIT Press, 2002), 29.

12. Louis Armand takes up a similar approach to argue not so much that "Joyce was necessarily in some way cognisant of a future possibility of hypertext" but that "Joyce's text can be said to *solicit* hypertext . . . the extent to which Joyce's text can be said to both *call for* and *motivate* a hypertextuality irreducible to a stable field." Armand and I differ, however, in that he is *not* interested in looking retrospectively at Joyce "from the position of current computing technologies." Retrospectively viewing earlier authors through the lens of current cultural practices is simply unavoidable—"current computing technologies" saturate our every thought, our very language—and so this fact should be instead acknowledged openly rather than sidestepped. Louis Armand, *Techné: James Joyce, Hypertext & Technology* (Prague: Univerzita Karlova v Praze, Nakladatelství Karolinum, 2003), xi.

13. Christopher Bantick, "Poetry's Death by a Thousand Hits," *Australian,* January 15, 2011, http://www.theaustralian.com.au/news/features/poetrys-death-by-a-thousand-hits/story-e6frg6z6-1225987952516.

14. Noah Wardrip-Fruin makes a similar plea for greater historicizing of terms such as *hypertext* that are often used to describe such digital poems. Similar to the literary-focused version of media archaeology I advocate in this chapter, he calls for an understanding of "the history of our terms so that we may see how competing definitions of the moment are movements in different directions from a common starting point." Noah Wardrip-Fruin, "What Is Hypertext?," Hyperfiction.org, http://www.hyperfiction.org/texts/whatHypertextIs.pdf, 1.

15. C. T. Funkhouser, *Prehistoric Digital Poetry: An Archaeology of Forms, 1959–1995* (Tuscaloosa: University of Alabama Press, 2007), 1.

16. Ibid., 3.

17. Ibid., 12.

18. Susan Howe wrote in 1985 that "Emily Dickinson and Gertrude Stein are clearly among the most innovative precursors of modernist poetry and prose." Eliot Weinberger reminds us, however, in his preface to the newly published New Directions edition of Howe's *My Emily Dickinson,* that not too long ago Alan Tate's assessment of Dickinson was not

particularly debatable—she is ignorant; she "cannot reason at all. She can only *see*." Not too long ago, Denise Levertov's early assessment of Dickinson in a letter to Robert Duncan was not particularly outrageous. In it Levertov admits saying to herself, "Jesus, what a bitchy little spinster." In fact, it has been only twenty-seven years since R. W. Franklin published *The Manuscript Books of Emily Dickinson,* thereby giving us access to a stunningly intelligent, erudite, experimental Dickinson. Susan Howe, *My Emily Dickinson* (New York: New Directions, 2007); Eliot Weinberger, preface to *My Emily Dickinson,* by Susan Howe (New York: New Directions, 2007), vii–viii.

19. This practice of self-conscious scholarship in the age of the digital is nicely paralleled by Martha Nell Smith's work with the *Dickinson Electronic Archives,* which is informed by what she calls a "technology of self-consciousness" that is as firmly positioned against black-boxing and in favor of processual knowledge as is this book: "Self-consciousness is a technology with which humanists are familiar. . . . The technology of self-consciousness required by computer encoding of texts produces a healthy self-consciousness about what Bruno Latour and Steve Woolgar describe in *Laboratory Life* as 'black-boxing'—which occurs when one 'renders items of knowledge distinct from the circumstances of their creation.' . . . Maintaining relentless self-consciousness about how critical 'facts' have been produced, about how items of knowledge are part of the circumstances of their creation, is crucial for responsibly providing the provisionality that characterizes the best kind of science of chaos." Martha Nell Smith, "Computing: What's American Literary Study Got to Do with IT?," *American Literature* 74, no. 4 (December 2002): 852–53.

20. Sharon Cameron, *Choosing Not Choosing: Dickinson's Fascicles* (Chicago: University of Chicago Press, 1992).

21. Smith, "Computing," 845.

22. Emily Dickinson, *The Manuscript Books of Emily Dickinson,* vol. 2, ed. R. W. Franklin (Cambridge, Mass.: Belknap Press / Harvard University Press, 1981).

23. Marjorie Perloff, "Screening the Page/Paging the Screen: Digital Poetics and the Differential Text," in *New Media Poetics: Contexts, Technotexts, and Theories,* ed. Adalaide Morris and Thomas Swiss (Cambridge, Mass.: MIT Press, 2006), 162.

24. Perloff, "Screening the Page," 162.

25. Steven Johnson, *Interface Culture: How New Technology Transforms the Way We Create and Communicate* (New York: Basic Books, 1997), 18.

26. Joel Spolsky, *User Interface Design for Programmers* (New York: Apress, 2001), 60.

27. Janet Murray, *Inventing the Medium: Principles of Interaction Design as a Cultural Practice* (Cambridge, Mass.: MIT Press, 2011), 10.

28. Ibid., 339.

29. Ibid., 292.

30. Cristanne Miller, "Whose Dickinson?," *American Literary History* 12, no. 1 (Spring/Summer 2000): 232.

31. Henry Petroski, *The Pencil: A History of Design and Circumstance* (New York: Alfred A. Knopf, 1989), 334.

32. Dickinson, *The Manuscript Books*, vol. 2, ix.

33. Han, "Jeff Han"; emphasis my own.

34. Lev Manovich, *The Language of New Media* (Cambridge, Mass.: MIT Press, 2001), 64–65. Amazon.com's ever more-popular Kindle is yet another example of the kind of contemporary, closed interface I discuss throughout—a device that Jeff Bezos, Amazon founder and CEO, first described as a "wireless, portable reading device with instant access to more than 90,000 books, blogs, magazines and newspapers." That number now exceeds one million, and further, in early 2011 Amazon reported that the sales of e-books exceeded those of paperback books. From one version to the next, the aim of the Kindle is to supplant the book. Reading the fine print of the license agreement and terms of use, however, you read, "You may not sell, rent, lease, distribute, broadcast, sublicense or otherwise assign any rights to the Digital Content or any portion of it to any third party, and you may not remove any proprietary notices or labels on the Digital Content." Or you see the warning against "Reverse Engineering, Decompilation, Disassembly or Circumvention." Or you find that all your reading and annotations will be monitored by Amazon: "The Device Software will provide Amazon with data about your Device and its interaction with the Service (such as available memory, up-time, log files and signal strength) and information related to the content on your Device and your use of it (such as automatic bookmarking of the last page read and content deletions from the Device)." So while Bezos might like you to "get lost in your reading and not in the technology," he is in fact asking you to see through the interface as if it were a neutral medium instead of a medium that prevented you from sharing, lending, and reselling these e-books; that disallowed you from engaging with the e-book as an art object; and that didn't imbed a layer of surveillance into your most private moments spent reading. "Amazon Kindle E-book Downloads Outsell Paperbacks," BBC News, http://www.bbc.co.uk/news/business-12305015.

35. Petroski, *The Pencil*, 334.

36. Susan Howe, "Some Notes on Visual Intentionality in Emily Dickinson," *HOW(ever)* 3, no. 4 (January 1987): http://www.asu.edu/pipercwcenter/how2journal/archive/print_archive/alertsvol3no4.html#some.

37. R. W. Franklin, introduction to *The Manuscript Books of Emily Dickinson*, vol. 1, by Emily Dickinson (Cambridge, Mass.: Belknap Press / Harvard University Press, 1981), 1413.

38. Emily Dickinson, "We met as Sparks," in *The Manuscript Books of Emily Dickinson*, vol. 2, ed. R. W. Franklin (Cambridge, Mass.: Belknap Press / Harvard University Press, 1981), 1052.

39. Walter Benn Michaels, *The Shape of the Signifier* (Princeton, N.J.: Princeton University Press, 2006), 5–6.

40. *Merriam-Webster's Collegiate Dictionary*, 11th ed., s.v. "spark"; emphasis my own.

41. Marta Werner, "The Flights of A 821: Dearchizing the Proceedings of a Birdsong," in *Voice, Text, Hypertext: Emerging Practices in Textual Studies*, ed. Raimonda Modiano, Leroy F. Searle, and Peter Shillingsburg (Seattle: University of Washington Press, 2004), 307, 308.

42. Franklin, introduction to *The Manuscript Books*, 848.

43. Jaishree Odin, "The Database, the Interface, and the Hypertext: A Reading of Strickland's *V*," *electronic book review*, October 14, 2007, http://www.electronicbookreview.com/thread/electropoetics/isomorphic.

44. While I do not discuss this aspect of A 92–14, surely the poem that is on the recto of "We met as Sparks," whose first line is "As one does sickness over," should also be attended to, as it cannot but affect/inform our reading, since each poem on each side of the sheet meets and departs from the other. For instance, even though the slip of paper is pinned to the verso, what should we make of the fact that we can also see the pin and the piercing made by the pin on the recto, which happens to be right beside the two variants "Habit" and "handle"?

45. Theodor Nelson, "Computer Lib / Dream Machines," *The New Media Reader*, ed. Noah Wardrip-Fruin and Nick Montfort (Cambridge, Mass.: MIT Press, 2003), 330.

46. George Landow, *Hypertext 2.0: The Convergence of Contemporary Critical Theory and Technology* (Baltimore: Johns Hopkins University Press, 1997), 3.

47. Mary Flanagan, "[the house]," in *The Electronic Literature Collection*, vol. 1, ed. N. Katherine Hayles, Nick Montfort, Scott Rettberg, and Stephanie Strickland, http://collection.eliterature.org/1/works/flanagan__thehouse.html.

48. Daniel C. Howe and Aya Karpinska, "open.ended," in *The Electronic Literature Collection*, vol. 1, ed. N. Katherine Hayles, Nick Montfort, Scott Rettberg, and Stephanie Strickland, http://collection.eliterature.org/1/works/howe_kaprinska__open_ended.html.

49. Ibid.

50. Judd Morrissey, "The Jew's Daughter," in *The Electronic Literature Collection*, vol. 1, ed. N. Katherine Hayles, Nick Montfort, Scott Rettberg, and Stephanie Strickland, http://collection.eliterature.org/1/works/morrissey__the_jews_daughter.html.

51. Ibid.

52. N. Katherine Hayles, "Intermediation: The Pursuit of a Vision," *New Literary History* 38, no. 1 (2007): 116.

53. Matthew Mirapaul, "Pushing Hypertext in New Directions," *New York Times*, July 27, 2000, http://partners.nytimes.com/library/tech/00/07/cyber/artsatlarge/27artsatlarge.html.

Postscript

1. "Updating Our Privacy Policies and Terms of Service," Google blog, January 24, 2012, http://googleblog.blogspot.com/2012/01/updating-our-privacy-policies-and-terms.html.

2. Félix Guattari, "The Best Capitalist Drug," in *Chaosophy: Texts and Interviews 1972–1977*, trans. Sylvère Lotringer (Los Angeles: Semiotext(e), 2009), 153.

3. John Battelle, *The Search: How Google and Its Rivals Rewrote the Rules of Business and Transformed Our Culture* (New York: Penguin, 2005), 4.

4. Ibid., 6.

5. "What It Does," Google website, accessed March 19, 2013, http://www.google.com/glass/start/what-it-does (emphasis my own). Google is even using Apple's own marketing against it, as Sergey Brin is quoted as asking at a Ted.com event on Google Glass: "Is this the way you're meant to interact with other people? It's kind of emasculating. Is this what you're meant to do with your body?"—implicitly both reminding us of Apple's penchant for feminizing its devices by describing them as thin, elegant, gorgeous, sleek and suggesting that Glass is somehow much more masculine. Rob Williams, "Google Co-founder Sergey Brin Feels 'Emasculated' by Smartphones," *Independent*, March 1, 2013, http://www.independent.co.uk/news/world/americas/google-cofounder-sergey-brin-feels-emasculated-by-smartphones-8514748.html.

6. Given how invested so many writers continue to be in a notion of the literary that they believe reflects their values of self-expression and originality, it is perhaps not surprising that it is easier for technologists than for poets to recognize not just Google's ubiquity but also its potential for harvesting poetic language. In their coauthored 2003 *Google Hacks*, the thoroughly nonliterary technologists Tara Calishain and Rael Dornfest propose "GooPoetry" to their readers. Preceding their inclusion of the code for the entirety of the GooPoetry CGI script, they tell us: "Google sure can mix a mean word salad. This hack takes a query and uses random words from the titles returned by the query to spit out a poem of random length. The user can specify a poetry 'flavor,' adding words to the array to be used." Tara Calishain and Rael Dornfest, *Google*

Hacks: 100 Industrial-Strength Tips & Tools (Sebastopol, Calif.: O'Reilly & Associates, 2003), 273.

7. "Software Is Mind Control—Get Some," I/O/D website, http://bak.spc.org/iod.

8. Geert Lovink, "Interview with the Makers of the *Web Stalker* Browser, Simon Pope, Colin Green and Matthew Fuller," I/O/D website, April 24, 1998, http://bak.spc.org/iod/nettime.html.

9. Geert Lovink, *Networks without a Cause: A Critique of Social Media* (Cambridge, UK: Polity Press, 2011), 152.

10. "Google Will Eat Itself," *GWEI* website, http://gwei.org/index.php.

11. Ibid.

12. *Google Gravity*, Mr. Doob website, http://www.mrdoob.com/projects/chromeexperiments/google-gravity.

13. *The Revolving Internet*, Constant Dullart website, http://therevolvinginternet.com.

14. Ibid.

15. Evgeny Morozov, "Keep Calm and Carry On . . . Buying," *New York Times*, March 9, 2013, http://www.nytimes.com/2013/03/10/opinion/sunday/morozov-the-surreal-side-of-endless-information.html?smid=tw-share&_r=2&.

16. Lev Manovich, *Software Takes Command* (New York: Bloomsbury Academic, 2013), http://lab.softwarestudies.com/p/softbook.html, 5.

17. Sol Lewitt, "Paragraphs on Conceptual Art," DDOOSS, http://ddooss.org/articulos/idiomas/Sol_Lewitt.htm.

18. Jasia Reichardt, ed. *Cybernetic Serendipity: The Computer and The Arts* (London: Studio International, 1968).

19. Ibid., 14.

20. Ibid., preface, unpaginated.

21. Archie Donald, "TIME-SHARING," in *Computer Poems,* ed. Richard Bailey (Drummond Island, Mich.: Potagannissing Press, 1973), 10–11.

22. Racter, *The Policeman's Beard Is Half Constructed: Computer Prose and Poetry*, ed. William Chamberlain and Thomas Etter (New York: Warner Software / Warner Books, 1984).

23. Racter, introduction, unpaginated; emphasis my own.

24. Charles Hartman and Hugh Kenner, *Sentences* (Los Angeles: Sun and Moon Press, 1995), 82.

25. Steve McLaughlin and Jim Carpenter, "Issue 1," in *Against Expression: An Anthology of Conceptual Writing,* ed. Craig Dworkin and Kenneth Goldsmith (Evanston, Ill.: Northwestern University Press, 2011), 414.

26. Ron Silliman, untitled blog post, *Silliman's Blog,* October 5, 2008, http://ronsilliman.blogspot.com/2008/10/one-advantage-of-e-books-is-that-you.html.

27. Marc Adrian, "Computer Texts," in *Cybernetic Serendipity,* 53.

28. Marshall McLuhan, "The Medium is the Message," in *Understanding Media: The Extensions of Man* (Cambridge, Mass.: MIT Press, 1994).

29. I should note that no matter how captivating, there are a number of works I do not touch on in this postscript that are referred to as either conceptual or flarf (a group of poets concerned with using search engines to produce a poetry of the inappropriate) by their authors and whose output text partially or entirely comes from search engine queries. For all their strengths, they do not seem particularly concerned with critiquing the workings of the search engine itself. For example, flarfist K. Silem Mohammad describes the process of creating his 2003 *Dear Head Nation* as one that involved punching "a keyword or keywords or phrase into Google and work[ing] directly with the result text that gets thrown up." Likewise, Katie Degentesh fed phrases from statements in the Minnesota Multiphasic Personality Inventory into various search engines and then pieced together poems she then published as *The Anger Scale* in 2006. K. Silem Mohammad, "Spooked and Considering How Spooky Deer Are," in *Against Expression: An Anthology of Conceptual Writing,* ed. Craig Dworkin and Kenneth Goldsmith (Evanston, Ill.: Northwestern University Press, 2010), 437; Katie Degentesh, *The Anger Scale* (Cumberland, R.I.: Combo Books, 2006).

30. Bill Kennedy and Darren Wershler, *apostrophe* (Toronto, ON: ECW Press, 2006).

31. Ibid., 293.

32. Ibid., 20.

33. Ibid., 51, 55.

34. Tan Lin, *Plagiarism/outsource : notes towards the definition of culture : untilted Heath Ledger project: a history of the search engine : disco OS* (La Laguna, Canary Islands: Zasterle Press, 2007).

35. Ibid., unpaginated.

36. Danny Snelson, "Heath, Prelude to Tracing the Actor as Network," *Aphasic Letters,* 2010, http://aphasic-letters.com/heath.

37. "What Is Copyleft?," Gnu Operating System, March 10, 2013, http://www.gnu.org/copyleft.

38. Tan Lin, *Plagiarism/outsource,* unpaginated.

39. Ibid., unpaginated.

40. Ibid., unpaginated.

41. In terms of its emphasis on reproduction, it's no surprise that in 2012 Lin published a new, revised version of *HEATH* titled *Heath Course Pak,* adding another fifty-two pages of new material, including an interview, autographed photographs of Jackie Chan and Heath Ledger, and numerous other modes of writing and writing technologies such as electronic post-it notes. Tan Lin, *Heath Course Pak* (Denver, Colo.: Counterpath Press, 2012).

42. John Cayley and Daniel C. Howe, *How It Is in Common Tongues* (Providence, R.I.: NLLF Press, 2012).

43. Ibid., 299.

44. John Cayley and Daniel C. Howe, "Common Tongues," ELMCIP Electronic Literature Knowledge Base, posted by Elisabeth Nesheim, http://elmcip.net/node/4677.

Index

Adrian, Marc, 176–77
algorithm, xiv, xxi, 65, 163–64, 166; search engine and, 164, 171–84. *See also* Google; Racter
AltaVista, 178–80
Apple (computers): Apple Lisa, xv, 61, 79, 80, 196; Apple II (-e, -plus, -c) and, xv, xix, 48, 53, 19, 64–67, 70–79, 195, 197; BASIC and, 31, 66–70; iOS and, 19, 24; iPad and, ix, x, 11, 13–19, 21, 23–30, 32, 49, 165, 188, 189; iPhone and, 11, 23, 49, 85, 188, 194; Macintosh and, xv, xix, 47, 48, 49, 50, 51, 52, 55, 61, 64–65, 71, 76–85, 165, 194, 198, 199; TV commercial and, 80–83
appliance: computers as, xi, 8, 77, 80, 142
archive, 132, 134, 137, 138, 179, 185. *See also* Media Archaeology Lab
artificial reality, 20. *See also* Krueger, Myron

Bailey, Richard, 174–75
Bernstein, Mark. *See* hypertext: Jane's Spaces and; hypertext: Storyspace and
bissett, bill, 101, 102–4, 106, 107, 110, 115, 121, 203
black box, xvii, 1, 3, 24, 32, 34, 37, 38, 55, 142, 212
bootstrapping. *See* Engelbart, Douglas
bpNichol. *See* Nichol, bp
Breeze, Mary-Anne. *See* Mez

Cage, John, 65, 191, 197
Carpenter, Jim, 175–76
Cayley, John, xiii, 136, 164, 166, 183–84
Chopin, Henri, 100, 101, 203
Chun, Wendy, 49, 52
code poem. *See* codework
codework, 4, 31–32, 69, 191
command-line interface, xv, xviii, xix, 6, 44, 46, 48, 49, 50, 63, 64, 77, 92, 194
Commodore, xv, 65, 77
conceptual writing, 38, 171, 217; Kenneth Goldsmith and, 163; Bill Kennedy and, 164, 171, 178–80; plagiarism and, 181, 182. *See also* Cayley, John; Lin, Tan; Wershler, Darren
concrete and dirty concrete poetry, 100–104, 117, 191, 203
Cook, Tim, 13, 17
creativity, xi, xvi, xvii, 5–6, 18–19, 27, 48, 64, 77–78, 83, 85, 189

Dada, 3, 137, 191, 203
defamiliarization, xviii, 2, 21, 30, 46; Victor Shklovsky and, 34–35
Dick, Philip K., 187
Dickinson, Emily, 129, 133–34, 135, 138, 139, 140, 142–43, 144, 152, 211–12; fascicle and, ix, xx, 145–50, 153, 154–55, 162
DIY, xv, 25, 31, 66, 70, 92, 94, 95, 98, 108, 123–27, 191, 192, 209
Donald, Archie, 174–75
dynabook. *See* Kay, Alan

Engelbart, Douglas, 47, 51–55, 58, 61, 63, 77, 194; demo and, 140
Ernst, Wolfgang. *See* media archaeology: Wolfgang Ernst and
expansion slots, 48, 64, 78–80, 194

Flanagan, Mary: "[theHouse]," 125, 135, 136, 152–55
Fluid User Interface (Fluid UI), 5
Foucault, Michel, 131–32, 185, 200, 201, 210
Fry, Ben, 123–24
Funkhouser, Christopher, 137–38
Futurism, 3, 127, 192

Galloway, Alexander, x, xii, 129, 131, 133
Gassée, Jean-Louis, 47, 48, 65
Gibson, William, 4, 36–38, 189
glitch, xi, xviii, 2; clean and, 40; digital literature and, 30, 35; glitch aesthetics and, 35–36, 191
Goldberg, Adele, 59, 60, 195. *See also* smalltalk
Goldsmith, Kenneth. *See* conceptual writing
Gomringer, Eugen, 99, 201, 203
Google: digital literature and, 178–80, 181–82, 215; Glass and, ix; x, 165–66, 215; googlization and, xxi, 163, 166, 170, 180, 182; net art and, 168–70; terms of service (TOS) and, 168–70. *See also* algorithm
Graphical User Interface (GUI), x, xvi, 48, 49, 50–52, 64, 77, 139. *See also* Apple (computers): Apple Lisa and; Apple (computers): Macintosh and; Xerox: Star and

hacking: digital literature and, 30, 32; McKenzie Wark and, 31, 50
Han, Jeff, x, 5, 131, 143–44

Hartman, Charles O., 174–75
Hayles, N. Katherine, 38, 133, 158
HCI. *See* Human Computer Interaction
Higgins, Dick, 89–90, 98
homebrew, 19, 31, 193, 209
Homebrew Computer Club. *See* homebrew
Houédard, Dom Sylvester, 87, 94–96, 98, 104, 106, 115–18, 125, 126, 191, 207
Howe, Daniel C., xiii, 136, 152, 155, 157, 164, 183–84
Human Computer Interaction, 51, 132, 133
Huth, Geof, 70–71
hypertext, 4, 134, 140, 150–52, 211; Jane's Spaces and, 33–34; Storyspace and, 32–34, 76

IBM, 58, 81, 104, 139, 197
ideology, xi–xii, xv, xvi, 2, 4, 49, 64, 80–84, 175. *See also* user-friendly
interface: definition of, x; interface-free and, x, xxi, 47, 29, 141–44, 148
invisible: interface and, ix, xi, xii, xiv, xvi, xvii, xviii, 1–6, 8–9, 11, 14, 16, 18, 21, 30–46, 49, 50, 63, 70, 85, 92, 122, 129, 130, 133, 141, 144, 163, 164, 177, 181, 183
Invisible Seattle, 197
Ive, Jony, 14, 15

Jobs, Steve, 14–15, 18, 48–49, 64, 79–80
Jodi, 36
Johnson, Ronald, 95, 97–98
Johnson, Steven, 3, 51, 141, 194
Joyce, James, 90, 211

Karpinska, Aya, and Daniel C. Howe, 135, 136, 152, 155–56, 157

Kay, Alan, 47, 51, 54–64, 77, 124, 125

Kenner, Hugh, 174–75

Kirschenbaum, Matthew, 37, 75–76, 186, 189

Kittler, Friedrich. *See* media archaeology: Friedrich Kittler and

Krueger, Myron, xvii, 19–22, 24, 27

Larsen, Deena, xiii, xviii, 4, 32–33

Levy, Steven, 14, 80, 84

Lewis, Jason Edward, 4, 23, 25–27, 28

Lin, Tan, xiii, 164, 180–82, 217

Logo. *See* Papert, Seymour

Lovink, Geert. *See* media archaeology: Geert Lovink and

Loyer, Erik, 4, 23, 27–30

Maeda, John, 123, 124

magic, ix, x, xvii, 4, 7, 10–11, 21, 34; iPad and, 13–19, 188

McCaffery, Steve, 87, 99, 107, 126, 203, 207; *Carnival,* 105, 114, 118–23, 125, 127, 202, 208, 209; Marshall McLuhan and, 104–5, 111, 121; typewriter and, 113–15

McLaughlin, Stephen, 175–76

McLuhan, Marshall: *Astronauts of Inner-Space,* 106, 117; concrete poetry and, 87, 98–99, 104–5; *Counterblast,* 207; *Culture Is Our Business,* 91; media archaeology and, 89, 91, 130; "The Medium Is the Message," 177; *Understanding Media,* 55, 89, 108; *Verbi-Voco-Visual Explorations,* 89–90, 200–201

media archaeology, 2, 49, 88–89, 121, 130, 131, 135, 164, 185; Wolfgang Ernst and, 132, 210; Friedrich Kittler and, xii, 2, 89, 185, 199; Geert Lovink and, xii,

166, 167; Jussi Parikka and, xiii; Siegfried Zielinski and, xiii, 1, 2, 87, 91–92

Media Archaeology Lab, xiv–xv, xvi, 186

media poetics, x, xiv, xvii, 50, 125, 129, 142, 163, 166, 171, 178, 182, 197

Mez, 4, 31–32; Mezangelle and, 32

Mezangelle. *See* Mez

mimeograph, 99, 106, 121, 205

Montfort, Nick, xviii, 4, 32, 186, 190, 191

Morrissey, Judd, xviii, 4, 34, 39, 43, 46, 135, 136, 152, 155–62, 202

multitouch, x, xvi, xvii, 2, 4, 5, 11, 14, 20, 27. *See also* Apple (computers): iPad and; Apple (computers): iPhone and; Han, Jeff

Natural User Interface (NUI), xvii, 1, 5, 6, 131

Nelson, Jason, xiii, xviii, 4, 34, 40, 41, 135, 202

Nelson, Theodor, 150–51, 153, 195

Nichol, bp, 87, 101, 119, 126, 203; "The Complete Works," 109; copier machine and, 110–12; *First Screening,* 66–70, 139; *Konfessions of an Elizabethan Fan Dancer,* 106–8; Marshall McLuhan and, 104–5, 110–11; *Translating Translating Apollinaire,* 109–11; typewriter and, 104–10, 113

NLS. *See* Engelbart, Douglas

Noigandres, 90, 99

obsolete, 38, 50, 75, 94, 197

Olson, Charles, 92–93, 99, 202

oN-Line System (NLS). *See* Engelbart, Douglas

open source, 31, 32, 95, 124, 125, 127, 172, 191

Organic User Interface (OUI), xvii, 1, 5, 7
Oulipo, 137, 191

Papert, Seymour, 47, 51, 53–54, 55, 57, 58, 61, 63, 77, 195
Parikka, Jussi. *See* media archaeology: Jussi Parikka and
Piringer, Jörg, xviii, 4, 23, 25, 26

Racter, 174
Raskin, Jeff, 79–80
readingwriting, xiv, xx, xxi, 163, 166, 170, 171, 177–78, 180, 182–84
Reas, Casey, 123–24
Reichardt, Jasia: *Cybernetic Serendipity,* vii, 172–73, 176
remediation, 135

Saroyan, Aram, 94, 108
Schiller, Phil, 11, 13
Scobie, Stephen, 101, 203, 204
seamless: interface and, 1, 3, 8, 34, 140, 165
search. *See* algorithm; AltaVista; Google
smalltalk, 54, 57–61, 196. *See also* Kay, Alan
Stefans, Brian Kim, 125, 126
Swigart, Rob, 2

transparency. *See* transparent
transparent: interface and, xi, xii, xiii, 39, 40, 77, 78, 102, 107, 122, 124, 126, 129, 142, 144, 152, 156, 162, 176, 182, 198
typestract. *See* Houédard, Dom Sylvester; McCaffery, Steve
typewriter, 50, 133, 193; carriage and, 123; interface and, 129, 130, 133; monospace and, 93, 95, 180. *See also* concrete and dirty concrete poetry

ubicomp. *See* ubiquitous
ubiquitous: computing and, xvi, xvii, 1, 2, 4–5, 6, 8–14, 16, 49, 85, 187
user-friendly, xi, xv, xvi, xviii, 2, 3, 4, 18, 34, 47, 49, 77, 79, 80, 85, 130, 131, 133, 164, 174, 193, 195. *See also* ideology

variantology, xiii, xiv, 88–89. *See also* media archaeology: Siegfried Zielinski and
Videoplace, xvii, 18–23, 25, 27. *See also* Krueger, Myron
Videotouch, 20–23. *See also* Krueger, Myron

Wark, McKenzie. *See* hacking
Web Stalker, 166–67
Weiser, Mark, 2, 4–5, 8–14, 16, 187
Wershler, Darren, 123, 164, 171, 178–80, 199, 208. *See also* conceptual writing
Williams, Emmett, 89, 93, 95, 96, 101, 115
Wozniak, Steve, 48, 51, 64. *See also* Apple (computers): Apple II (-e, -plus, -c) and

Xerox: Alto and, 54, 59, 61, 79; Parc and, 9, 11, 57, 58, 79, 124; Star and, 61–64, 65, 79, 139, 196

Young-Hae Chang Heavy Industries (YHCHI), xiii, xviii, 4, 34, 40–46, 124

Zelevansky, Paul, xiii, xix, 50, 71–76; *The Case for the Burial of Ancestors,* 71–72; "SWALLOWS," 72–76
Zielinski, Siegfried. *See* media archaeology: Siegfried Zielinski and

Lori Emerson is assistant professor of English at the University of Colorado at Boulder. She is coeditor, with Marie-Laure Ryan and Benjamin J. Robertson, of *The Johns Hopkins Guide to Digital Media*; coeditor, with Derek Beaulieu, of *Writing Surfaces: Selected Fiction of John Riddell*; and coeditor, with Darren Wershler, of *The Alphabet Game: A bpNichol Reader*.

(continued from page ii)

29 *Games of Empire: Global Capitalism and Video Games*
Nick Dyer-Witheford and Greig de Peuter

28 *Tactical Media*
Rita Raley

27 *Reticulations: Jean-Luc Nancy and the Networks of the Political*
Philip Armstrong

26 *Digital Baroque: New Media Art and Cinematic Folds*
Timothy Murray

25 *Ex-foliations: Reading Machines and the Upgrade Path*
Terry Harpold

24 *Digitize This Book! The Politics of New Media, or*
Why We Need Open Access Now
Gary Hall

23 *Digitizing Race: Visual Cultures of the Internet*
Lisa Nakamura

22 *Small Tech: The Culture of Digital Tools*
Byron Hawk, David M. Rieder, and Ollie Oviedo, Editors

21 *The Exploit: A Theory of Networks*
Alexander R. Galloway and Eugene Thacker

20 *Database Aesthetics: Art in the Age of Information Overflow*
Victoria Vesna, Editor

19 *Cyberspaces of Everyday Life*
Mark Nunes

18 *Gaming: Essays on Algorithmic Culture*
Alexander R. Galloway

17 *Avatars of Story*
Marie-Laure Ryan

16 *Wireless Writing in the Age of Marconi*
Timothy C. Campbell

15 *Electronic Monuments*
Gregory L. Ulmer

14 *Lara Croft: Cyber Heroine*
Astrid Deuber-Mankowsky

13 *The Souls of Cyberfolk: Posthumanism as Vernacular Theory*
Thomas Foster

12 *Déjà Vu: Aberrations of Cultural Memory*
Peter Krapp

11   *Biomedia*
     Eugene Thacker

10   *Avatar Bodies: A Tantra for Posthumanism*
     Ann Weinstone

9   *Connected, or What It Means to Live in the Network Society*
     Steven Shaviro

8   *Cognitive Fictions*
     Joseph Tabbi

7   *Cybering Democracy: Public Space and the Internet*
     Diana Saco

6   *Writings*
     Vilém Flusser

5   *Bodies in Technology*
     Don Ihde

4   *Cyberculture*
     Pierre Lévy

3   *What's the Matter with the Internet?*
     Mark Poster

2   *High Technē: Art and Technology from the Machine Aesthetic
     to the Posthuman*
     R. L. Rutsky

1   *Digital Sensations: Space, Identity, and Embodiment
     in Virtual Reality*
     Ken Hillis